THE

MEASURE OF THE CIRCLE.

PERFECTED IN JANUARY, 1845,

BY

JOHN DAVIS.

PROVIDENCE:
PUBLISHED FOR THE AUTHOR.
1854.

THE MEASURE OF THE CIRCLE.

THE USE AND IMPORTANCE OF THE MEASURE, DISCOVERED IN JANUARY, 1845.

THE globe is divided into 360 degrees; each degree into 69½ miles; how can it be known what the 360th part is? A degree is a 360th part; a mile is a 69½ part of a degree; a yard is a 1760th part of a mile; a foot is a third part of a yard; an inch is a twelfth part of a foot; a barleycorn is a third part of an inch. Now, these are parts which are all unknown, and have been since the world began, as may be supposed.

I therefore say that all measures are imperfect, and, without a perfect quadrature of the circle, must remain so. I find the table of long measure is all imperfect, and so it is with all measurements, of all descriptions. There is nothing in shape to represent the noble works of Supremacy, known to man, that can be mathematically measured: such as land, if in a circular form; casks containing liquids; steamboat boilers; a grindstone. These cannot, by mathematics, be correctly measured.

Being encouraged to make this discovery by the large offers and appropriations by all governments, I devoted my best endeavors to effect a perfect quadrature; and have, as I believe, to the satisfaction of all who may examine my work, and to my own, beyond all possible doubt. As I have learned, this measure has been the strife and anxiety of all the most learned men that have lived since the world began; and, in defiance of all, it has slept in oblivion until the year 1845, when discovered, as is represented in this work.

I will set forth some of the objections that were made to me, to prove the impossibility of this measure, by different persons. A Mr. Clifford, the ninth wrangler of Cambridge University, said it was an utter impossibility, for it was a true figure of eternity; and, as eternity never could be measured, no more could a circle. My reply was, "I admit it to be a true figure of eternity, as there is neither beginning nor end to it; but there is a great difference in measuring it, for I can get at either side of a circle, and neither side of eternity." The gentleman replied, "Well, that does make a material difference." A Professor Olmsted said he thought it to be an impossibility, for it had been tried hundreds and thousands of years. My reply was, that it must be in nature, because I could get wrong upon the extreme, either way; and there must be a point between those two extremes that must be right.

And I say, in confidence, I have found the point.

I was advised to introduce my work at Cambridge, in England. I did so; and was told that the impossibility was so great that a man might as well try to shoot the moon.

I was recommended to G. B. Arry, professor royal at Greenwich. I called on this gentleman. He replied to me that it was not worth while for me to puzzle my brains about the measure of the circle, for it was measured as near as it ever could be; and he did not see what use it was, or ever could be, to them.

Now, it seems to me that this gentleman must be in an error, and grossly mistaken; for I cannot see how he can pretend to hold out to the people a complete traverse of the stars, without the knowledge of the measure of the circle, unless he may think it as well to teach the world a falsehood as the truth. His situation in life ought to warrant to the people better ability, and different expressions.

I now come to some facts as to the use and principles.

It is often said, in the construction of a steam engine, that the power is not equal to the calculation. They say the cause is, the steam does not properly condense; and they allow fifteen pounds' weight of air to each and every square inch. Now, what can be expected, when they know not how many inches their cylinder contains?

My experience as a mechanic has taught me the use and importance of this measure. I have been engaged and employed as a superintendent, having the care and sole management of cotton, linen, woollen, and silk machinery, and have gained the approbation of my employers, my management being such as to save to them, yearly, a large amount. They said they could not comprehend how I could manage to such perfection; for when I had perfected a thing, it was sure to answer the purpose. This perfection I arrived at by the measure of the circle, unknown to them.

The expression was made use of to one of my employers, in a cotton manufacturing village, "How is it that it does not cost you one half to keep your machinery in order that it does the rest of us?" The answer was, "When Davis calculates, it is sure to come right; while others have to do their work five or six times over."

I was called on to measure a circular stair rail. The gentleman wanted to know if I could calculate, by figures, the exact length of the rail. My reply was, that I could, if he gave me the height of the well. He gave me the height and diameter of his well. I gave him the length of the rail. He said I was wrong; it was but so long. "Well," said I to him, "then your rail is that much too short." "It is," he said; "I have tried it." This I consider to be a demonstration by mathematics that cannot be performed except by the perfect quadrature of the circle.

It is said that Sir Isaac Newton's opinion with respect to the mechanical powers was, that whatsoever was gained in time was lost in power; and whatsoever was gained in power was lost in time. This maxim does not hold good in all cases. Sir Isaac was well aware that gravitation was not known; neither can it be, without the measure of the circle; which when perfected, Sir Isaac's maxims are not correct, on account of the want of this measure — the long-sought solution of the perfect quadrature of the circle.

With these facts and circumstances, I shall leave to whoever may be interested in my welfare and interest to do for me what conscience may dictate to an honest heart. I ask or claim nothing but what good reasoning may, with sound judgment, honestly demand. I think I know the value and importance of this work, as my forefathers have, in all ages of the world. Utterance refuses expression, to paint in letters the utility and magnitude of this measure, which has been so long sought for.

It appears, through and by the wisdom of God, that this circular principle is what he has put forth in his wonderful creation of all things; and why this measure has slept in oblivion, unknown to man, is known only to God.

This measure will and must prove a great benefit to mankind, when understood, as it is the basis and foundation of mathematical operations; for, without a perfect quadrature of the circle, measures, weights, &c., must still remain hidden and unrevealed facts, which are and will be of great importance to rising generations. The improvements that will arise from this measure fifty years hence I cannot paint in imagination.

THE PRINCIPAL RULES.

I think that, after people become satisfied that my work is right, it will be but one hour's labor for the scholar to learn all that is necessary in practice. I will, before I lay down the work in the book, lay down the principal rules for the learner, that he may see it in the commencement as well as in the last pages of my book.

The use of the measure of the circle is to find the circumference of any circle, great or small, in order to correct and make right all weights and measurements, which are wrong, and have been since the world began.

The proportion the diameter has to the circumference is as 6 to 19. The difference of as 7 to 22, or as 6 to 19, is as $\frac{1}{132}$ is to $\frac{1}{133}$. This makes linear measure $75\frac{3}{4}$, square measure $1.52\frac{1}{2}$, cubic measure 2.29 hundredths per cent. astray. This makes the foot rule near $\frac{1}{10}$ of an inch too short; the yardstick near $\frac{3}{10}$ too short.

A mathematical inch is the 38th part of a circle 12 inches in diameter.

Rule 1*st.* To find the circumference of any circle, great or small. — Multiply the diameter by $9\frac{5}{10}$, (this is my ratio, derived from as 6 to 19,) and divide the product by 3; this gives you the perfect circumference, in all cases. Suppose your circle is 12 inches in diameter: —

```
      12
       9.5
     ---
     108
       6
     ---
 3 ) 114
     ---
      38, circumference.
```

Rule 2*d.* To find the area of the same circle. — Take 3 times the radius, by once the radius; this gives the square inches. Divide the square inches by 3 raised to its 4th power, or biquadrate. Then add the 4th power, or biquadrate, to the square. This gives the perfect area, in all cases. Suppose the diameter is 12 inches, radius 6: —

```
       6
       3
     ---
      18
       6
     ---
 3 ) 108
     ---
 3 )  36
     ---                 Add to 108
 3 )  12                          4
     ---                        ---
       4, power.                112, area.
```

Rule 3d. — Or you may take $\frac{7}{9}$ of the square of the diameter; that will give the perfect area, in all cases. Suppose the diameter is 12 inches: —

```
     12
     12
   ----
    144
      7
   ----
9 ) 1008
   ----
    112, area.
```

Rule 4th. Having the circumference, to find the diameter. — Divide the circumference by 19, and multiply the quotient by 6, which gives the diameter, in all cases. Suppose the circumference is 38: —

```
19 ) 38 ( 2
     38   6
     --  --
      0  12, diameter.
```

Rule 5th. — Let the cooper take 3 times the diameter, with one third the radius; his head will just fill.

Rule 6th. — Cut a thin strip of brass or tin, — the thinner the better, — 38 inches long, form a perfect circle, and the diameter is 12 inches.

Rule 7th. Having the area, to find the circle that bounds it. — Suppose any number of figures as an area; for instance, let it be 448; divide by 28; subtract the quotient from the given sum; divide the remainder by 3; then extract the square root, which will be the radius of the circle that bounds the figures. The square root of any number of figures operated on this way will be the radius of the circle, in all cases.

```
     144 ( 12, √     28 ) 448 ( 16       448
     1                    28              16
     ---                  ---            ---
22 ) 044                  168        3 ) 432
     ---                  168            ---
      44                  ---            144
     ---                    0
       0         Radius, 12 × 2 = 24.
```

MEASURE OF THE CIRCLE.

Time is prefigurative of the number 6; so, as time is measured by a circle, I take that number to measure a circle.

Suppose a circle to be 12 inches in diameter. I take my radius, and multiply 12, diameter, by ratio $9\frac{5}{10}$; the product is 114; divide this by 3, which gives 38, the circumference.

Proof.—I begin with a hexagon, each side and radius being of equal length, say six inches. 6 multiplied by 6 is equal to 36; this is the sum of the 6 sides of the hexagon. I call every polygon 36 inches, when it is 12 inches in diameter. I begin with a figure having 4 sides, it being easy to understand. Take a strip of paper 36 inches long, and $\frac{1}{4}$ of an inch wide; cut it into strips of 9 inches each, and place them so as to form a square; the four corners will then be vacant. To supply these, I take $\frac{1}{6}$ of the radius, which is equal to one inch, and, for variety, call it the 7th. So I square the 7th by the number of sides. Multiply 4 by 4, = 16, square $\frac{1}{4}$. I add 1 for each angle, to fill the vacant corners, which lengthens the square from 9 to $9\frac{1}{2}$ inches. $9\frac{1}{2}$ multiplied by 4 is 38, which is the circumference, when curved to a circle, gaining 2 inches. As every polygon will gain 2 inches, if worked right, I add 2 to 36, which gives 38, circumference.

I now begin with 6 sides and square the 7th, thus: 6 multiplied by 6 is equal to 36; square sixth coned thirds. Add 1 for each side, which is equal to 2; 2 added to 36 is equal to 38, the circumference.

I now take 8 sides, and square the 7th, thus: $8 \times 8 = 64$, the 8th coned, $\frac{1}{4}$. Add $\frac{1}{4}$, equal side, = 2. 2 added to 36 is 38, the circumference.

I take 10 sides, and square the 7th, thus: $10 \times 10 = 100$, tenth coned fifth, $\frac{1}{5}$ for each side. 2 added to 36 equal 38, the circumference.

I take 12 sides, and square the 7th, thus: $12 \times 12 = 144$, twelfth coned sixth, $\frac{1}{6}$ each side. 2 added to 36 equals 38, the circumference.

This is to prove that the diameter is to the circumference as 6 to 19. Now, I take 6 for my diameter, and suppose it to be formed into a hexagon. The 6 sides equal 18 inches, and 18 multiplied by 18 is equal to 324. Add the square of the diameter, 36, to 324, = 360, the number of degrees in a circle. Now, the square root of 360 is equal to $18\frac{36}{37}$, or $\frac{1}{18}$; as there are no corners wanting to a circle, it leaves 19 inches for the measure of the circle. The circumference of the circle is 19 inches; but the square root of 360 lacks $\frac{1}{37}$; so you see my diameter is 6 inches, and my circumference is 19, which makes it as 6 to 19. The measure of the circle is but linear measure; therefore 36 makes an inch on a line, in this case, as is now used. Now, in extracting the square root of 360, it lacks 1 of filling the square which is superfluous in a circle, as no corners are wanting.

I will now show how I came by my ratio.

I take the 19, and divide it by 6, and when divided, it will come to 3 whole numbers, and $166\frac{2}{3}$, decimal. Now, multiply $3.166\frac{2}{3}$ by 3. 3 times $\frac{2}{3}$ is 2 whole numbers in decimals. I multiply by 3, in order to bring it to whole numbers, so as not to use fractions; and, as I multiply this product by 3, I must, after multiplying the diameter by the ratio, divide by 3.

```
     12
    9.5
   ----
    108
      6
   ----
3 ) 114
   ----
     38, the circumference.
```

THOMAS JEFFERSON.

I find, in a work published by B. L. Raynor, in New York, in 1832, entitled "The Report of Thomas Jefferson to Congress, on Coins, Weights, and Measures, in 1790," that Mr. Jefferson says, in relation to all weights and measures, that "all are imperfect under the present system." The report from the secretary of state, containing a plan for a uniform system of coins, weights, and measures, on page 311 of this book, was executed with most astonishing despatch, considering the intricacy of the subject, and novelty of the experiment. In sketching the principles of his system, Mr. Jefferson was dependent on the guide of his own genius, as no example to dictate or direct his researches existed. It is somewhat remarkable that two of the principal governments of Europe were also engaged at this period on the same subject.

The first object that presented itself to his inquiries was the discovery of some measure of invariable length, as a standard. There exists not in nature, as far as has been hitherto observed, a single object, accessible to man, that presents one uniform dimension.

The globe of the earth might be considered as invariable in all its dimensions, and that its circumference would furnish an invariable measure; but no one of its circles, great or small, is accessible to admeasurement, in all its parts; and the various trials to measure different portions of them have resulted in showing that no dependence can be placed on such operations, for a certainty. Matter, then, by its mere extension, furnishes nothing invariable. Its motion is the only remaining resource.

The motion of the earth on its axis, though not absolutely uniform and invariable, may be considered as such, for all human purposes. It is measured, obviously, but unequally, by its departure from a given meridian of the sun, and its return to that meridian, constituting a solar day. Throwing together the inequalities of solar days, a mean interval, or solar day, has been found, and divided by general consent into 86,400 parts, called seconds of time.

Such a pendulum, then, becomes itself a measure, of determined length, to which all others may be referred, as a standard. But even the pendulum was not without its uncertainty, as the period of its vibration varied in different climates or latitudes. To obviate this objection, he proposed the standard might refer to a particular latitude; and that of 38 degrees being the mean latitude of the United States, he adopted it.

THE CIRCLE.

The circle is one of the noblest representations of Deity, in his noble works of human nature. It bounds, determines, governs, and dictates space, bounds latitude and longitude, refers to the sun, moon, and all the planets, in direction, brings to the mind thoughts of eternity, and concentrates the mind to imagine for itself the distance and space it comprehends. It rectifies all boundaries; it is the key to information of the knowledge of God; it points to each and every part of God's noble work; it divides east from west, north from south, with all its variations so beautiful; it brings to the thoughts of man the eternity and incomprehensibility of that space and distance which none can explore or determine. Nothing but imagination can cope with its extent, bounded as it is by the most extreme, unseen and unsought distance that minds can imagine. It has neither beginning nor end; its bounds are unknown; its area cannot be told by numbers; no mouth can reveal its magnitude; it is what contains all human flesh and blood; it contains all the improvements susceptible to the ingenuity, science, and activity of the human family; it brings to mind that eternity of bliss and happiness where the weary are at rest, and from whence no traveller returns; it contains monuments of marble and stone, in memory of those who have sought its quadrature in all ages of this world; it is the companion of every man, woman, and child; it enters the families of all the earth; it is the mediator of honesty, harmony, and content, in all; it rectifies those principles which are calculated to comfort and console honesty to the bosom of all. Its want of correctness has been the strife and anxiety of all the learned, and the lovers of science, since the world began. The wrongs which man has done to man, for want of a perfect measure, are numerous. Immense sums of money, with much time and anxiety, have been spent by all nations for its perfection; and no one nation, perhaps, is worthy of more respect for its exertions than France.

Its proportion is now found. It can be measured to perfection. It can no longer slumber, for its equality can now be expressed.

DIAGRAMS OF CIRCLE.

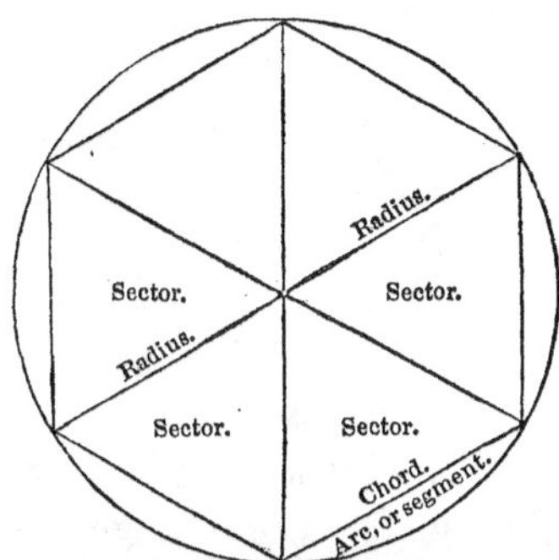

This is a hexagon, having 6 equal sides.

This is to describe the four strips of paper set forth in the work, supposed to be 9 inches square, leaving the vacant corners as described.

The hexagon is what I measure the circle by, it being the only figure known by which the circle can be measured, and the number 6 the only number.

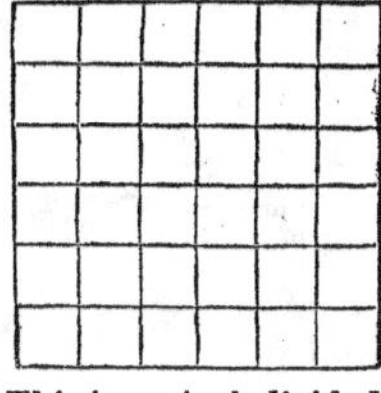

This is an inch divided into sixths.

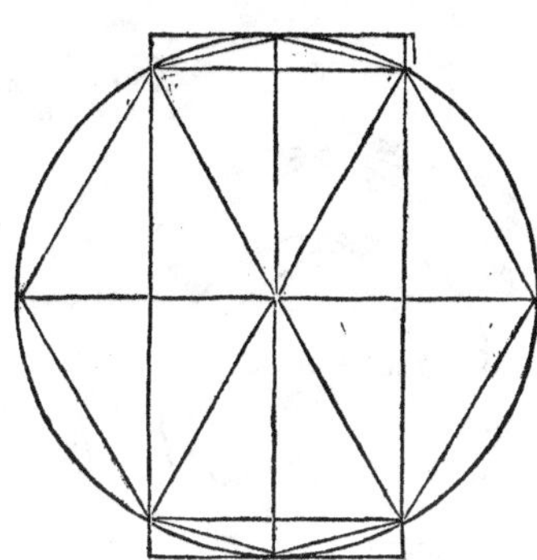

This is an oblong square, 6 by 3 inches.

```
       18
        6
     ----
 3 ) 108
     ----
  3 ) 36
     ----
  3 ) 12
     ----
        4
```

This is the 4th power, or biquadrate. Add this to the square, and it gives the area of the circle.

```
12, diameter.  108
                 4
               ---
               112
```

The question has been asked by many, why some other figure will not answer as well as the hexagon. The hexagon is equal in all its parts, and no other figure is. Had I taken any other, it would have carried me into surds, which would be beyond comprehension.

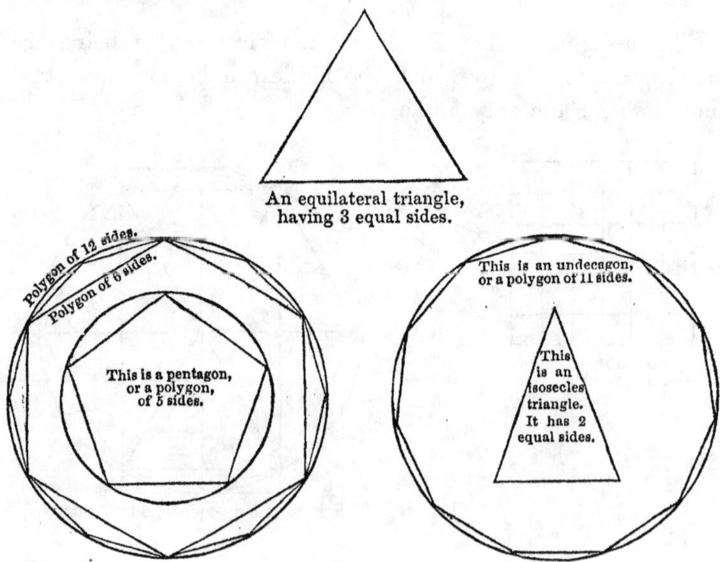

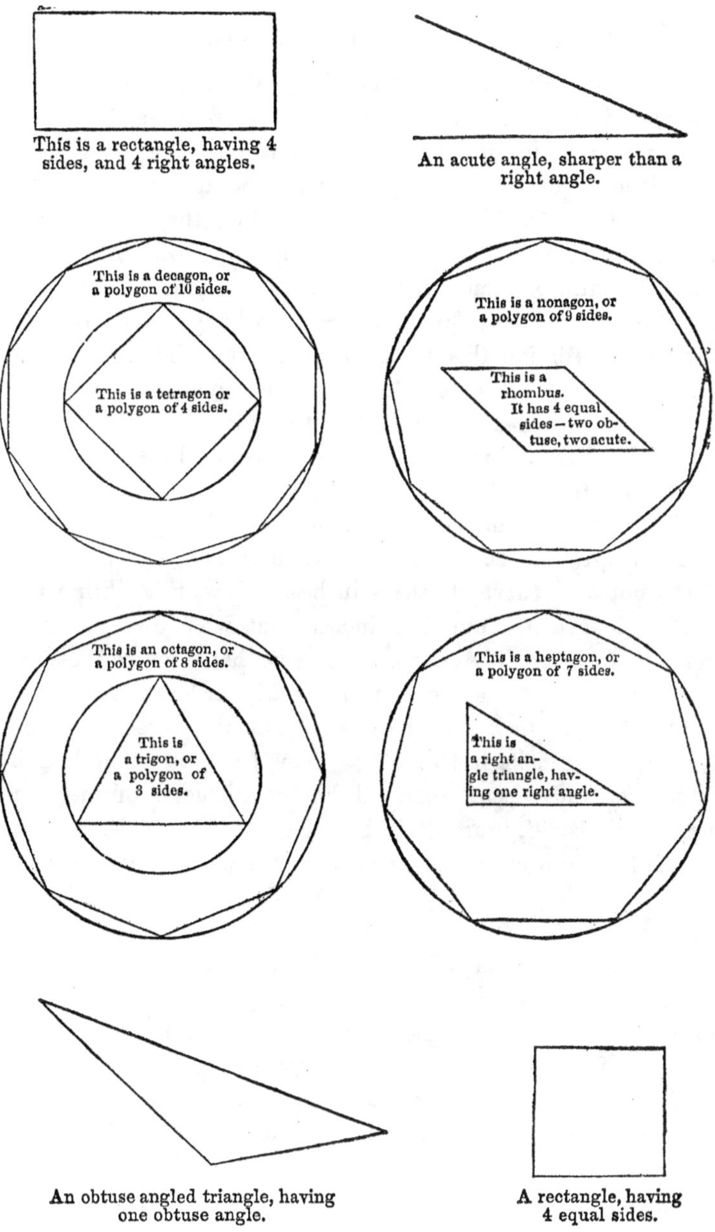

This is a rectangle, having 4 sides, and 4 right angles.

An acute angle, sharper than a right angle.

An obtuse angled triangle, having one obtuse angle.

A rectangle, having 4 equal sides.

TO MEASURE A CIRCLE BY RINGS.

Suppose the diagram is a triangle, $\frac{1}{6}$ of an inch scale. To measure this angle, I divide the circle into 36 rings, each ring $\frac{1}{6}$ of an inch wide, and take the measure of the 18th ring, which is as follows: Take 112, which is the area of the circle, 12 inches being the diameter. Now, to find this area, I take the square of the radius, which is 6 inches, and multiply it by 6, $= 36$; multiply this 36 by 3, $= 108$; divide this 108 by 3, $= 36$; divide this 36 by 3, $= 12$, which is the cube, or third power; divide this 12 by 3, $= 4$, which is the biquadrate; then I add the biquadrate to the square, 108, and it gives the perfect area, thus: $108 + 4 = 112$, area of circle.

Now I multiply this area, 112, by 6, which throws the area into a strip 672 inches long, and $\frac{1}{6}$ of an inch wide; this I divide by 36, the number of rings in the circle; this gives the average length of each ring, on a straight line, which is $18\frac{4}{6}$ inches, but on a curve it is 19 inches. Now, this 18th ring, is, when lengthened out, $18\frac{4}{6}$ inches, but when brought to a circle, 19 inches. The reason for this is, that when you change this straight line to a circle, you take, in order to make the concave, $\frac{2}{6}$ of an inch from the inside, which goes to the length of the ring, to make it 19 inches. Now I will show the length of the first and second ring. Take the diameter of the first ring, which is 36, multiply it by 9.5, $= 342$; divide this by 3, $= 114$, length of the first ring. For the second ring I take 35, and multiply it by 9.5, $= 110.833\frac{1}{3}$. Now, in order to make the work shorter, I take the 18th ring, as being the average length of the 36 rings, which I have shown above, and multiply its measure, 19 inches, by the radius, 6 inches, thus: $19 \times 6 = 114$ inches, area of the angle, but not of the circle, as I shall prove. The circumference of the outside ring is 38 inches; multiply this by 3, which gives 114. Second ring, $110.833\frac{1}{3}$, which is just $3\frac{1}{6}$ less, equal to $1\frac{1}{18}$ on a circle, but only 1 on a straight line. So the triangle measures $\frac{1}{18}$, which is equal to $\frac{2}{36}$ too long. I therefore take twice 36, which is equal to 2 inches, from 114, which leaves 112, area of circle.

This is a diagram one sixth of an inch scale.

This is the measure of the 36 rings, $\frac{1}{6}$ of an inch scale.

114.
110.833⅓
107.666⅔
104.5
101.333⅓
98.166⅔
95.0
91.833⅓
88.666⅔
85.5
82.333⅓
79.166⅔
76.0
72.833⅓
69.666⅔
66.5
63.333⅓
60.166⅔
57.0
53.833⅓
50.666⅔
47.5
44.333⅓
41.166⅔
38.0
34.833⅓
31.666⅔
28.5
25.333⅓
22.166⅔
19.0
15.833⅓
12.666⅔
9.5
6.333⅓
3.166⅔

Each and every one of these rings loses in length one inch and $\frac{1}{18}$. Every ring is one inch and $\frac{1}{18}$ shorter inside than on the outside.

Time being prefigurative of the number 6, I take that number to measure the circle. Suppose my radius is 6 inches, by

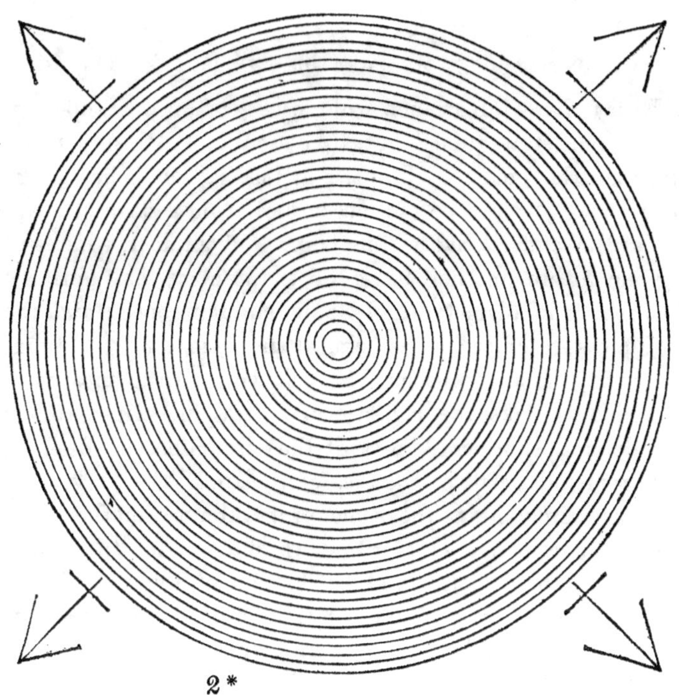

$\frac{1}{6}$ of an inch; then my circle will be 12 inches, which, brought to a hexagon, will measure 36 inches, equal to the square of the radius. Now, I divide one side of the hexagon into 36 squares, and, supposing my radius fixed by one corner to a point at the centre of the circle, I move the other end just its width, which is $\frac{1}{6}$ of an inch on a straight line, which gives the biquadrate, equal to $\frac{1}{36}$ of the radius; and, as it doubles on a circle, it is equal to $\frac{1}{18}$ gained on the side of the hexagon, equal to $\frac{2}{36}$, which added to 36 degrees, make 38, the circumference, or $6\frac{1}{3}$ inches for $\frac{1}{6}$ of the circle.

So 6 prefigures time; time, with its square, with the biquadrate conically added, measures the circle; therefore, to 3 times the diameter, equal 36, add the square root of the biquadrate, equal 2, which gives the circumference. Also, to 3 times the square of the radius, 108, add the biquadrate, 4, which gives 112, the perfect area of the circle.

OBSERVATIONS ON THE MEASURE OF THE CIRCLE BY RINGS.

1. Every ring $\frac{1}{6}$ of an inch wide loses $\frac{1}{3}$ of an inch, upon an average, when straightened out, by reason of the inside being shorter than the outside. Suppose a ring be 19 inches long; 18 inches and $\frac{4}{6}$ would be the measure of it on a straight strip, square at the ends. Every ring being $\frac{1}{6}$ of an inch wide, and fitting one within another, being 36 in number, they must be just 12 inches in diameter, and 38 in circumference; and the average measure of each ring being 3 and $\frac{1}{9}$, and 36 the number of rings, the area must be 112. If you wish to measure a ring separately, take the measure of a straight strip $\frac{1}{6}$ of an inch wide, and subtract $\frac{1}{57}$ part of the circumference from it, and the remainder will be the area of the ring.

2. Every circle bears the same proportion to another as a square. As, suppose a square 6 inches, and another 12 inches; the square of the 6 inch would be 36, and of the 12, 144, or 4 times as much; and if it was 24 it would be 16 times as much. So, in a circle 6 inches in diameter the area is 28 inches; but 12 is 4 times 28, equal to 112; and 24 is 16

times as much, or 448, the area, although the circle would only double, as, 6 diameter, 19 circumference, and 12 diameter, 38, &c. A line to go round a circle would be just in the same proportion.

3. To find the measure of the circle. Suppose 12 is the diameter; then the radius will be 6. Now find the biquadrate, thus:—

$$\begin{array}{r} 6 \\ 6 \\ \hline 36 \\ 6 \\ \hline 216 \\ 6 \\ \hline 1296 \end{array}$$

1296, the biquadrate, or square of the square, as $36 \times 36 = 1296$.

Now take a hexagon, divide one side into 36 sixths of an inch, and the radius $\frac{1}{6}$ of an inch wide. Cut one side so that you can move it its width to its centre point. The outside angle being on the hexagon, by moving the centre $\frac{1}{6}$ to the centre point, you raise $\frac{1}{6}$ from the next angle; but as it must meet the angle again it gains double the biquadrate, just the same as the square; so that instead of gaining $\frac{1}{36}$, it gains $\frac{1}{18}$ of the hexagon, or any part of it. And this is the perfect measure; for as it proceeds from its own centre, there cannot be any defect in the measure, any more than in the circle itself; for at the same time that it gains $\frac{1}{36}$ on a straight line, it gains $\frac{1}{36}$ the other way, which when curved will be just $\frac{1}{18}$; and 36 sixths on a straight line, and twice on a curve, equal 38 sixths, one side of the hexagon, just 38 inches the whole circle.

4. The square of a circle 12 inches in diameter is $9\frac{1}{2}$ inches; this is the square of the circle, but is not the square of the area of the same circle. The square of the area of a circle 12 inches in diameter is about 10.583 inches.

The fourth part of a circle is the square of the circle itself, but not the area. The circle gains over all other measures. Take a circle 38 inches round, the diameter of which is 12 inches, and the area of this circle is 112 inches. Now, take a square 38 inches round, that is, 9½ inches on each of its four sides, and the area will be 90¼ inches. The area of a circle 12 inches in diameter will be 112 inches; and of a circle 6 inches in diameter, 28 inches. So you see that it varies the same as the square. Half the diameter is ¼ in area; also, 3 times the diameter gains the biquadrate in circumference, and also its square in area; as, $2 \times 2 = 4$, area.

Now, this is strong proof of the perfect measure of the circle, and beyond a doubt; because as the square varies, so varies the area; they are synonymous; one proves the other, in all cases.

5. From Euclid. The impossibility of expressing the exact proportion of the diameter of a circle to its circumference by any received way of notation has put the most celebrated men in all ages upon approximating the truth as near as possible, there being a necessity of a nearer quadrature, inasmuch as it is the basis upon which the most useful branches of mathematics are built.

The famous Von Keulen carried it to 36 places of decimals, which he ordered to be placed on his tombstone, thinking he had set bounds to further improvements. The next that attempted it with success was the indefatigable Mr. Abraham Sharp, who, by a double computation, viz., from the sine of 6 degrees one way, and from the sine and cosine of 12 degrees another way, carried it to twice the number of places that Von Keulen had done, viz., 72. Afterwards, Professor Machin, by different methods of computation, carried it to 100 places.

Now, it is spoken of as a most wonderful feat of exactness, of scientific mathematical knowledge, to perfect a complete measure of the circle; and I have heard learned men say, "I have carried it to 50 and 100 places." Archimedes did not suppose that he had found a perfect ratio or an exact proportion of the diameter to the circumference of a circle, when he said it was nearly as 3 times, or 7 to 22.

What probability is there to a mathematical mind that by running the figures out — if it were possible — it would effect a perfect measure of the circle, as long as the exact proportion of the diameter to the circumference was not found?

In order to make a ratio that will measure a circle correctly, you must first have the exact proportion the diameter has to the circumference, as the ratio is for a common multiplier, to find the circumference, and is to be derived from the exact proportion the diameter has to the circumference of a circle. I have found, by the operation of figures, that this proportion is as 6 to 19. Now, in order to make a ratio, I divide the 19 by 6, which gives 3.166$\frac{4}{6}$. I make a vulgar fraction of $\frac{4}{6}$, and reduce it to $\frac{2}{3}$. I find the value of 3.166$\frac{2}{3}$ by multiplying by 3, which brings it into whole numbers, 9$\frac{1}{2}$. This as a ratio will answer for whole numbers in all cases, because it finds the substance of both dividend and divisor. My proportion of the diameter to the circumference is perfect, and that being perfect, makes my ratio perfect. But any ratio to be derived from as 7 to 22 would be imperfect, inasmuch as the proportion is imperfect. The ratio to find the area of a circle is 7854. Reason must dictate to all that square measure is not circular, because an angle of 90 degrees is square measure, and an angle of 60 degrees is circular.

To perfect the exact area of any circle by square measure, which is what this ratio is derived from, is impossible, from the nature of things, as by this ratio it makes the area of a circle 12 inches in diameter to be 113$\frac{1}{10}$ inches. This is just square measure; and circular measure of the same circle is 112 mathematical inches. Now, you see these angles differ 30 degrees. See the following demonstration: 12 being the diameter, they make the circumference 37$\frac{7}{10}$, and the area 113$\frac{1}{10}$. Suppose them divided into 36 rings; divide the circumference thus: $37.7 \div 2 = 18.85$, mean length of rings. Now, multiply by the radius, 6 inches, which will give 113$\frac{1}{10}$ inches, area of the rings, which is exact square measure, instead of circular.

By mathematical attention to these facts you will see that no square measure can perfect circular, which is what measures the circle; and you will find it impossible to perfect the measure of the circle by square measure.

In a circle 12 inches in diameter form a hexagon and a dodecagon, and take one of these angles, and the angle of the hexagon and dodecagon will measure just 18 inches the 6 angles; the 6 angles added together will be 108 inches. Now, the segment of this circle will measure just 4 inches, which is the 4th power, or biquadrate, and makes up the measure of the circle. That is its area; 3 times the diameter, with the biquadrate added, is the perfect circumference. Also, 3 times the square of the radius, with the biquadrate added, is the perfect area.

Dividing by 3 to the 4th power, is the same as dividing by 27. It is the $\frac{1}{27}$ part added to the square; as $3 \times 3 = 9$, and $9 \times 3 = 27$. The biquadrate is $\frac{1}{18}$ of 3 times the diameter of the circle; and the biquadrate is also $\frac{1}{27}$ part of 3 times the square of the radius, which makes up the area.

6. To convince that 90 degrees is a wrong angle to measure a circle, and to find the true area of a circle 12 inches in diameter. The largest ring in the plate of the 36 rings is supposed to be 38 inches in circumference, and the 18th half the size, or 19 inches. Now, the way they measure is to multiply half the circumference, equal 19, by the radius: thus, $19 \times 6 = 114$. By this you measure every ring by square measure, that is, 90 degrees. Thus,

19 inches long, and 1-6 wide.

equal 3 1-6 inches.

Multiply 36 rings, the number of rings in the circle: —

$$
\begin{array}{r}
36 \\
3\frac{1}{6} \\
\hline
108 \\
6 \\
\hline
114 \text{ inches.}
\end{array}
$$

But a circle falls short of the measure of a straight strip, as the inside is shorter than the outside. So proceed thus: take the hexagon 6 inches, as laid down in the work; the circle is 19 inches.

19 long, and 1-6 wide.

Set a bevel to 60 degrees, and place it on the outside of the hexagon, and you will find it falls short of 90 by 30, equal to $\frac{1}{3}$ to each angle; so the 6 angles lose $\frac{2}{36}$ of an inch; and 19 inches being the average length of the rings, each ring loses $\frac{2}{36}$ of an inch, which multiplied by 36 are equal to 2 inches, to be subtracted from 114, leaving 112, the perfect area of the circle.

This is the measure of the area of a circle 12 inches in diameter; and if the area is right, the circumference by which the area is bounded must be right also, for one proves the other.

7. To show the error of the old trigonometrical way of measuring the circle: Take 12 for diameter; suppose it to be made into 36 rings, as before shown; then the longest ring is 38 inches on a line, on the outside, and every ring is $\frac{1}{18}$ shorter on the inside than on the outside; so the 36 rings lose just $\frac{36}{18}$ of an inch, equal to 2 inches from 114 — leaving 112 for true area; but in their way the 12 inch ring, instead of 38 circumference, is but $37\frac{7}{10}$. Now, divide $37\frac{7}{10}$ by 2, equal 18.85, the mean length of rings. Divide them into 36 straight strips 18.85 long, $\frac{1}{6}$ of an inch wide, square at the ends, equal to 6 inches wide; so 18.85 multiplied by 6, equals $113\frac{1}{10}$, area; but if made into 36 rings, their area would be but $111\frac{1}{10}$.

8. To prove that the present measure in use, that makes the circumference of a circle 12 inches in diameter to be $37\frac{7}{10}$, is perfect square measure, because it makes the area of this circle $113\frac{1}{10}$ inches, I draw a diagram, and suppose it to be $37\frac{7}{10}$ inches long, and 6 inches radius. I make two angles, which will be 6 inches at one end, and a point at the other, and one of these angles will be just half the oblong square, which is $113\frac{1}{10}$ inches, according to the measure in use.

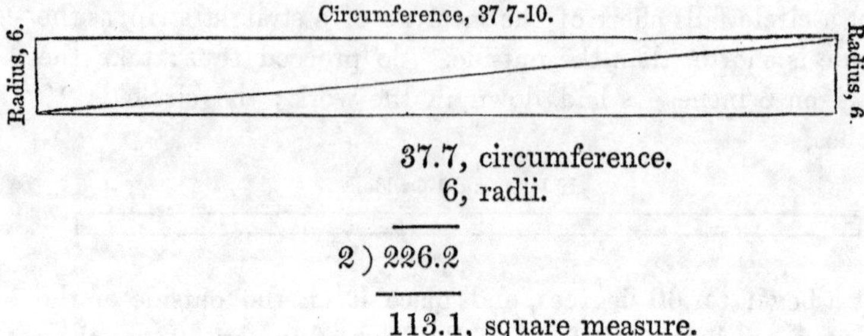


```
     37.7, circumference.
        6, radii.
     ----
2 ) 226.2
    -----
    113.1, square measure.
```

To show the common error of the measure now in use, which makes $113\frac{1}{10}$ the area of a circle whose diameter is 12, and circumference $37\frac{7}{10}$, which area is $1\frac{1}{10} = \frac{11}{10}$ too much, I will perform an operation by tenths.

```
38 ) 110 ( 28 36/38        28           36
     76                     2            2
     ---                   ---          ---
     340                   56      2 )  72
     304                    1           ---
     ---                   ---          38
      36                   57 34/38     34
                                        ---
                                        38
```

This shows that if the circumference is $37\frac{7}{10}$, and the area $113\frac{1}{10}$, the diameter should be nearly $12.057\frac{34}{38}$, instead of 12.

A RULE FOR GAUGING.

Having effected the measure of the circle, and knowing that circular measure has never been in use, and believing that the gauging of casks must be imperfect, many complaints being made of the present gauging, I have investigated the subject, and found the true and mathematical measure by the circle.

By specific gravity, $62\frac{1}{2}$ pounds of distilled water are 1 cubic foot; and the law provides that a gallon shall weigh 8 pounds of distilled water, weighed at the level of the sea. I find that 221 inches weigh 8 pounds, according to the statute gallon; I also find the circular inches in a gallon to be 284. Now, in order to effect this measure, I procured a box nearly 6 inches square, and $6\frac{1}{2}$ inches in depth; then I boiled some Croton water, and strained it through a flannel cloth; I then poured into the vessel until it weighed 8 pounds, and found it filled the vessel $6\frac{1}{8}$ inches, as near as I could determine by the eye, which I consider as perfect measure as could be obtained from such measuring.

By mathematics I find that a vessel 6 inches square, in order to hold a gallon, must be 6 inches and $\frac{1}{8}$, and $\frac{1}{72}$ part, in depth; and I consider the $\frac{1}{72}$ part as not perceptible to the eye. I therefore come to the conclusion that the gallon is precisely 221 inches, as I know it to be by mathematics, thus:—

```
  6
  6
 ——
 36
  6
 ——
216 inches, lacking 5 of a gallon.
  5 I add the 5 inches.
 ——
221, the square inches in a gallon.
```

Now, $\frac{1}{8}$ of 36 is $4\frac{1}{2}$; and $\frac{1}{72}$ of $36 = \frac{1}{2} = 5$, which added to $216 = 221$ inches, or 8 pounds.

To find the circular inches in a gallon. The circle is $\frac{7}{9}$ of the square; therefore 221 is $\frac{7}{9}$ of 284, which is $\frac{9}{9}$, thus:—

```
7 ) 221
    ———
    31 4/7
     2
    ——
    63
   221
   ———
   284, the number of circular inches
            in a gallon.
```

Rule for measuring the contents of a cask. — Take the diameter of the head and bung, and add them together; take $\frac{1}{2}$ of their sum, which will give the mean diameter. Multiply the mean diameter by itself, and that product by the length of the cask; then divide the last product by 284, and the quotient will be the answer, in gallons and parts.

For example: Multiply 31.5 by 28.7 = 3261636. Divide this by 284 = $114\frac{1}{2}\frac{80}{84}$ gallons.

The more to convince, I have obtained from one of the gauging masters fourteen casks, and compared them with my measure, and find the gauging to be from 1 to 5 per cent. more than my measure, which is circular. I have tried my measure by actual experiments, by measuring water into casks, and found it to be as perfect as can possibly be expected from such measuring.

No. of cask.	Length.	Bung.	Head.	Gallons.	True measure.	Difference.
1	36.9	32.3	29.3	121	117	1
2	35.7	33.5	30.0	126	126	0
3	37.3	29.3	26.0	102	100	2
4	36.9	31.5	28.3	117	116	1
5	35.0	33.9	29.4	124	123	1
6	37.6	33.4	29.2	129	130	1
7	37.9	29.9	26.5	107	106	1
8	49.9	31.6	28.5	130	130	0
9	37.8	33.3	30.0	134	133	1
10	38.8	33.7	29.2	135	131	4
11	35.4	33.3	29.6	125	123	2
12	40.0	32.0	24.0	115	110	5
13	35.1	32.3	29.5	118	118	0
14	36.8	32.7	28.8	124	123	1

The above table represents the gauging of 14 casks, by sworn gaugers in New York city, as handed to me. I find, in this number of casks, that the gaugers make 20 gallons more than I do by my rule of gauging; so that I think it will vary from 1 to 5 per cent. too much.

To show the measure of a circle by rings. Suppose 12 is the diameter; $12 \times 9.5 = 114 \div 3 = 38$, the circumference. Now, the longest ring is 38 inches long, and $\frac{1}{6}$ of an inch wide.

Suppose 36 straight strips $\frac{1}{6}$ of an inch wide and 38 inches long would measure just the same as a piece of board 38 inches long and 6 inches wide.

$$\begin{array}{r} 38 \\ 6 \\ \hline 2\,)\,228 \\ \hline 114 \end{array}$$

Dividing 228 by 2 would give 114, if the rings were all straight strips. But these strips are the same measure on both sides, while the inside of the rings is $\frac{1}{36}$ shorter, reducing the measure of the strips $\frac{1}{57}$, equal to 2 inches; therefore subtract 2 from 114, leaving 112 for the rings.

To prove my measure of the circle by the square, draw a square just 6 inches in diameter; then 6 multiplied by 6 will be 36, the area; then draw another of 12 inches, which is 2 diameter; but 12 multiplied by 12 is 144, just 4 times the area. So twice the diameter gives 4 times the area; and whatever is the result in a square will be the same in proportion in a circle, if it is measured right.

So you may see in my work that 6 for a diameter gives 28 for the area, which is just 4 times the same, in proportion, as the square, which proves the measure correct, beyond a doubt; for if the area is correct, it is impossible for the circle that bounds it to be otherwise.

I have been so much troubled in trying to convince people of the measure of the circle, and its area, that I am almost prone to think that if some college-bred man were to say twice 2 was equal to 5, many of them would not appeal to common sense, but conclude at once that it was so. For instance, a rectangle of 90 degrees is commonly called a right angle; and I allow it to be a right angle for a square, but not for a circle. Its result would be 33⅓ per cent. astray.

The right angle for a circle is 60 degrees. But they tell me that 60 is not a right angle; and I know that 90 degrees is not a right angle for a circle. It would seem that they wish me to act the monkey or mimic, so that if they are wrong, I must be.

Now to the point. A circle 12 inches in diameter is 38 inches in circumference, as is proved in my work. By trigonometry, which is in the proportion of 90 degrees, I take the circumference for one side of the triangle; the extreme width is 6 inches, and the mean width is 3 inches; so I multiply 38 by 3, which gives 114 for the area.

I now come to circular measure. Take 6 times the measure of the oblong square, as laid down in this work, or 3 times the square of the radius, and either of them is just the measure of the dodecagon, or polygon of 12 sides, equal to 108, square measure.

Now to measure the circular part. The 6 sides of the hexagon measure 36 inches, and the circumference is 38, gaining 2 inches, which is $\frac{1}{18}$. I will work it, to show the difference:

If 90 : $\frac{1}{18}$: : 60
90
———
60) 1620 ($\frac{1}{27}$, answer.
120
———
420
420
———
0

If $\frac{1}{18}$: 6 : : $\frac{1}{27}$
18
———
27) 108 (4, answer.
108
———
0

4 added to 108 equals 112, area of circle.

Now, 112 is circular measure, and 114 is square measure, and in the proportion of 90 degrees; so you may divide 112 by 36, which will give the mean area of each ring, or you may measure them separately; and if the rings are measured right it must be the true measure of the circle; and its area for the one will prove the other.

The following diagram represents $\frac{1}{18}$ of a circle; so I cut off the circular part, and divide the remainder into 3 parts, equal in length to the chord of the circle. Set your square on the chord, and move the other end to the centre point, and it will leave an angle on the chord equal to $\frac{1}{18}$.

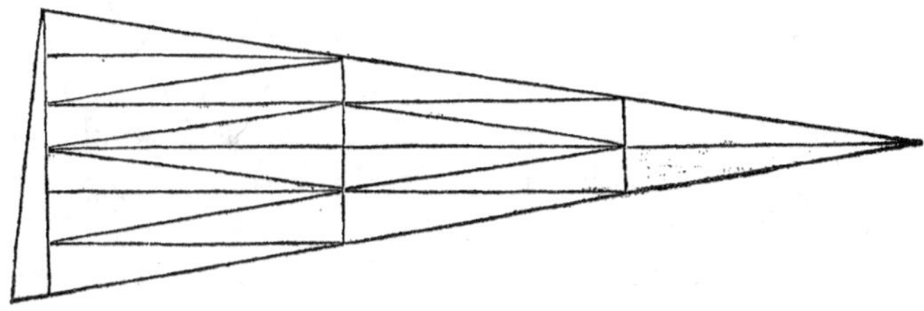

THE CIRCLE.

A circle is a line continued equidistant from a common centre till it ends where it began. The diameter is the line drawn from one side of the circle, through its centre, to the opposite side. The radius is the length of a line drawn from the centre to the circumference.

I think the most easy way for people to understand my measure of the circle is by the rings, which I have laid down in this work; of the correctness of which I think I could convince any man that would listen to common sense.

To find the area of a circle, divide the circle into six angles, or points, by the 6 radii, and take an oblong square from one of them, as before shown. Multiply that square by 6, and the product will be just the area of the dodecagon, or polygon of 12 sides. Find the biquadrate by dividing the area by 3 to its 4th power, and add the quotient to the area, and it will be the area of the circle, or space contained within the circle.

With regard to measuring a circle by polygons, with the help of mathematics, it has been disputed, by saying the polygons varied, and that I have called them 36 inches, which I allow to be so; for if one side of the hexagon is 6 inches, with 6 sides, the side of a polygon which has 12 sides would be 3 inches, to be in proportion. But it is $3\frac{21224}{2161339}$. This dispute did not arise from ignorance, as I think reason must dictate that I should prefer rational numbers to work with, when they answer the best purpose. Now, we know that the square and the hexagon are both perfect figures; and if the result of these is right, the result of all the others must be so, when worked the same way. And that they are so may be proved by the work; as, whatever polygon you try, it gives the same measure of the circle; and if it did not it would prove itself wrong.

This dispute seems as unreasonable to me as it would for a man to dispute that twice 10 made 20, because they are not counted one by one.

To find the circumference of a circle 6 inches in diameter: The 6 radii are 3 inches; so I take the square:—

```
       18
       18
       ——
       144
       18
       ——
       324, square of 18.
        36, the square of the diameter.
       ——
       360 (18 36/37
       1
       ——
  28 ) 260
       224
       ——
        36
```

So it wants $\frac{1}{36}$ of $\frac{1}{36}$ of being 19 inches. But 1 inch on a straight line is equal to $1\frac{1}{18}$ on a circle. So, when the $\frac{1}{18}$ is made to cover the two sides, which is always the case in the square root, it will be but $\frac{1}{36}$, which leaves the circumference just 19 inches.

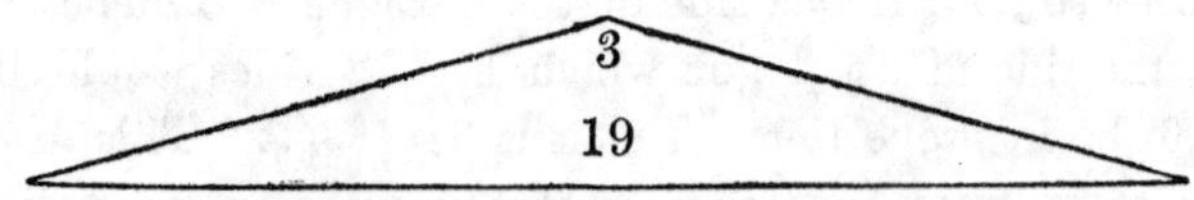

The area is 28.5 by trigonometry, as is used in the common way; but circular measure makes it only 28.

THE DIFFERENT RATIOS.

To show the measure of a circle 12 inches in diameter: Suppose you draw a circle 12 inches in diameter with a pair of compasses, from which you draw a hexagon, one side of which will be 6 inches on a straight line, but on curving it to a circle it will be found, by dividing it into 36 degrees, instead of 60, to bring the fractions perfect, which fractions are equal, to double the biquadrate found.

Thus the radius, 6 inches, $6 \times 6 = 36$ sixths of an inch, and one degree is $\frac{1}{6}$ of an inch, which added to $36 = 37$, the hypotenuse, $\frac{1}{36}$ of $\frac{1}{36}$ on a straight line; therefore in one side of the hexagon I gain just one degree, equal $\frac{1}{6}$ of an inch, and on a circle two degrees, equal $\frac{1}{3}$ of an inch, as the circle always gains double the same, in proportion, as the square. Although the circle does not make the square abruptly, it does make it, or has to make it; so the measure is mathematically true.

Diameter, 12 inches; 6 times 38 degrees, equal $\frac{1}{6}$ each — $38 \times 6 = 228 \div 6 = 38$ inches, circumference.

I will now measure a circle 12 inches in diameter by the measure now in use: —

$$\begin{array}{r} 3.1416 \\ 12 \\ \hline 37.6992 \end{array}$$

This makes it 37 inches, $\frac{6}{10}$, $\frac{9}{100}$, and $\frac{7}{1000}$.

I now find the area of the above by the measure now in use: —

$$\begin{array}{r} 7854 \\ 144 \\ \hline 31416 \\ 31416 \\ 7854 \\ \hline 13.0976 \end{array}$$

This is 113 inches, $\frac{0}{10}$, $\frac{9}{100}$, and $\frac{76}{1000}$.

I now find the circumference by my ratio, 12 being the diameter: —

$$\begin{array}{r} 12 \\ 9.5 \\ \hline 108 \\ 6 \\ \hline 3\,)\,114 \\ \hline 38, \text{ circumference.} \end{array}$$

I now find the area of the same; diameter 12 inches, radius 6 inches: —

$$\begin{array}{r} 6 \\ 3 \\ \hline 18 \\ 6 \\ \hline 3\,)\,108 \\ \hline 3\,)\,36 \\ \hline 3\,)\,12 \\ \hline 4, \text{ 4th power, or biquadrate.} \end{array}$$

Add the 4th power to the square, $108 + 4 = 112$, area of circle.

The difference of ratio 3.1416 and ratio 9.5 is: Ratio 9.5, 38 inches; ratio 3.1416, 37 inches, $\frac{6}{10}$, $\frac{9}{100}$, and $\frac{7}{1000}$; difference, 1.0976 inches; difference of area or circumference of same, $\frac{3}{10}$ of an inch.

So you see the measure now in use gives less circumference and more area, which is wrong, beyond all doubt.

ARCHIMEDES' MEASURE.

The proper ratio to be derived from as 7 to 22 is $3.1428\frac{4}{7}$, which measures a circle nearer than any measure that has been in use since Archimedes left it. All the attempted approximations have been farther off from the correct measure, instead, as has been thought, of being nearer. It has been a general prevailing idea that Archimedes' measure, as 7 to 22, was too much; and all have been striving to make it less.

I now find the area by rings, $\frac{1}{6}$ of an inch wide. If you begin at the centre point, and lay ring upon ring, you may not perceive it; but if you begin at the outside you will find 112 inches just meet the centre, which proves it right; whereas, 113.1 will run $6\frac{6}{10}$ inches beyond the centre, which proves it wrong, in spite of every thing that can be brought in its favor. Now, in a circle 12 inches in diameter, they make the circumference 37.7, and the area 113.1; I make the circumference 38 inches, and the area 112. So they make $1\frac{1}{10}$ inch more area, and $\frac{3}{10}$ less circumference.

I should like to know what to do with the $1\frac{1}{10}$ inches in the area of a circle 12 inches in diameter, and call it 38 inches in circumference, much more call it $37\frac{7}{10}$.

People seem determined to believe that the circumference is 37.7; but all can see by this that the rings would of course be 5.4 inches shorter, and would not reach the centre point by the 5.4 inches. But according to the area 113.1, they would run $6\frac{3}{5}$ beyond the centre, being that much too long; but by shortening the circumference and increasing the area, as has been done, the rings would run 12 inches beyond the centre point of the circle. But if the circumference and area were right, the rings would just strike the centre.

I find the square root of a circle 12 inches in diameter: —

```
         112 ( 10.58 636/2108
         1
         ----
  205 ) 01200
          1025
         -----
 2108 ) 17500
          16864
         -----
            636   The answer is 10.58 636/2108.
```

OF THE DEGREE.

If a degree is $69\frac{1}{2}$ miles, the earth's diameter is $7960\frac{20}{22}$: —

```
        360
        69½
      -----
       3240
      2160
        180
      -----
22 ) 25020 ( 7960 20/22
       7
     ------
     175140
     154
     ---
      211
      198
      ---
      134
      132
      ---
        2
```

As it is not convenient for me to find the diameter of the earth by astronomy, I have to take it from books, and they vary; but by the plan I have taken I make it 7906 English miles, which is 1 mile more than Hodson makes it; so I call it right. Now, by Archimedes' proportion, as 7 to 22, it makes the diameter $7960\frac{2}{22}$.

```
I make the circumference, 25020
              Diameter,   7901
                          -----
                          25020
                        225180
                       175140
                       ---------
                       197683020, square miles.
```

Mr. Hodson makes his

circumference, 24840

7900

22356000
173880

196236000

Now, I subtract Mr. Hodson's from mine: —

197683020
196236000

1447020

In squaring the circle I find in all cases, after measuring the 6 oblong squares, equal to 3 times the square of the radius, which is all that may be expected. The remainder is the biquadrate, which is termed segments, which in 12 inches diameter is just 4 inches, which added to the square inches, 108, which is 3 times the square of the radius, completes the area, 112. It is my opinion that, as in a round table the biquadrate measures the circle, or the circular part, it is when squared what is not improperly called squaring the circle.

The square root of 4 is 2; therefore the square of the circle 12 inches in diameter and 38 in circumference is 2 inches, which added to 3 times the square of the radius, or 3 times the diameter, makes 38, circumference. But the square of the area is $10.583\frac{4}{212}$.

THE DISPUTE.

My measure of the circle has been disputed, by saying my way would not be right in any number but 12; for by taking half the biquadrate the area would not be the same. This I agree to; but I never said take half the biquadrate, but take the contents of the biquadrate, which in 12 diameter would be 4, and add it to 3 times the square of the radius, and it would be the complete measure of the area, equal 112; and take the square root of the biquadrate, equal 2, and add it to 3 times the diameter, and it would complete the measure of the circle.

I am aware that though the square root of 4 is 2, equal to half the contents, the square root of 16 would be but 4, equal to one fourth the contents.

The straight line of a circle, or 6 sides of the hexagon, measures just 3 times the diameter, so that the circumference of 12 diameter would be 36 inches. But in curving that line to a circle, every sixth of an inch gains $\frac{1}{36}$ on a straight line, and twice that on a circle, which is equal to $\frac{2}{36}$, as proved by mathematics; so that 6 inches on a straight line is equal to $6\frac{1}{3}$ on a circle; but no instrument made with hands can demonstrate this measure.

I find, by examining different works, that people try to find the area of the circle by trigonometry, which is very imperfect, for its result is square measure, the angle of which is in the proportion of 90 degrees, whereas the angle of a circle is in the proportion of 60 degrees.

If 90 : $\frac{1}{18}$: : 60, answer $\frac{1}{27}$. Now, in a circle 12 inches in diameter, and whose radius is 6 inches, 3 times the square of the radius equals 108; divide 108 by 27, equal 4, which added to 108 equals 112, area.

But by trigonometry, if 90 : $\frac{1}{18}$: : 60, it gives $\frac{1}{12}$ for angle; and $\frac{1}{2}$ of 12 is 6, which added to 108 equals 114, area; making the segment that you cut off from the oblong square near $\frac{3}{4}$ of the square itself, whereas it should be but $\frac{1}{6}$; for there is not room in the circle to contain more. So you may see by the above that the circle gains $\frac{1}{18}$, which in 12 diameter is just 2 inches; for as one side of the hexagon is 6 inches on a straight line, 6 times 6 are 36, which divided by 18 equals 2 inches, which I gain by curving the straight line to a circle. So 36 and 2 added equal 38, circumference.

So the circle gains 2, and the square of 2 is 4, the square of the circle, independent of the dodecagon, or polygon of 12 sides. And 4 is also the 27th part of 3 times the square of the radius, equal to 108 inches, as seen above. And 4 is equal to the square of the biquadrate, thus: 3) 108 ; 3) 36 ; 3) 12 (4. So if I add $\frac{1}{27}$ to the straight lines of the rings, I make just the right area for them; and, consequently, the measure of the circle must be right, because the area will just fill the circle.

LEARNED MEN.

I have been told by great men that the measure of the circle was found as near as it ever could be, and it was near enough for any thing; but I find that, by the present measure in use, the convex surface of the earth falls short of the true measure 1.160107 geographical square miles. Dividing 1473336004 by 127 produces for answer 1.160107. Circumference, 21600; multiplied by 6821 produces 147333600. So it seems to me that correct measure will do no harm, for it is a good principle to know and practise the truth.

To prove that no measure is perfect without the perfect quadrature of the circle, I take the table of long measure, and prove by the great circle of the earth, which is so many miles, yards, feet, &c., that the carpenter's rule of 12 inches is near $\frac{1}{10}$ of an inch too short, and the yardstick near $\frac{3}{10}$ too short, and some square measure is still further astray. But some of our great men say it is near enough for any thing. Astronomers may pronounce it such, for if they are 10 or 20 millions astray, the vulgar know no better than to believe them; but if a mechanic made his joints lack as 7 to 22, we should have large rat holes in our buildings, and some of our learned men might be eaten up.

I received a letter from a professor of high standing, stating that it was discovered by mathematicians that if the diameter was 1, the circumference was 3.14159; so I work by that, and prove the difference, thus:—

$$\begin{array}{r} 3.14159 \\ 12 \\ \hline 37.69908 \end{array}$$

4

My circumference is 38; so I substract from

$$\begin{array}{r} 3800000 \\ 3769908 \\ \hline .30092 \end{array}$$ too little.

So you see it is $\frac{3}{10}$ and $\frac{92}{100000}$ too short in a circle 12 inches in diameter. To find what part it is too short in measure, divide the circumference by the difference: 30092) 3800000 (122956; divide the circumference by 122956:—

$$\begin{array}{r} 122956)\,3800000\,(\,3 \\ 368958 \\ \hline 11.132 \end{array}$$

So you see it is very near $\frac{3}{10}$ to a yard short of measure.

A circular staircase builder of New Haven — Mr. Bodsford — made a new circular stair rail, in the proportion of 1 to 3.14159; and as I was passing, he asked me if I could tell him the length of that rail by figures. I said yes, and gave him the length. He said, "You are wrong; it is only so long." I said, "Then your rail is that much too short." "Yes," said he; "it is about that much too short." So, instead of the rail coming over the floor at the landing, it was $1\frac{1}{2}$ inches over the well hole.

I have been told by mechanics that they had discovered that 38 was the circumference of 12 diameter.

SQUARING THE CIRCLE.

To square the circle, in the first place adopt some certain diameter, say 12 inches. Now, I know that an angle of 90 degrees is commonly called a right angle, but I think it might as well be called an erect angle, because it is erect from the horizon. It is a right angle for a square, but 45 is the right angle for a mitre, and any number of degrees would be the right angle for something.

The reason I make this statement is, people tell me there is no right angle but that of 90 degrees, and that I must call it so, or otherwise they could not understand me. But I call 60 degrees the right angle to measure a circle, and the only right one. Now, the groundwork to measure a circle is, 6 triangles equal to 18 perfect angles; and there is no number but 60 that will make them perfect; and if they are imperfect, my measure will be imperfect $\frac{1}{3}$ of $\frac{1}{6}$, equal to $\frac{1}{18}$.

In the first place, to the 6 sides add $\frac{1}{18}$, equal 2, in proportion of 90, linear measure, for the circumference. But $\frac{1}{18}$, although right in the linear, is wrong in the artificial. Now, 6 inches being the radius, $\frac{1}{18}$ is equal to 2, and the square of 2 is 4, which is the multiplier, and 2, equal $\frac{1}{2}$ the angle, is the divisor. Now, there are 18 angles, of 60 degrees each; therefore $\frac{1}{3}$ of $\frac{1}{18}$ equals 54, divided by 2 equals 27. Now add the 4, which in the proportion of 90 degrees is $\frac{1}{18}$, but in the proportion of 60 is $\frac{1}{27}$, thus:—

$$\begin{array}{r} 27 \\ 1 \\ \hline 28 \\ 4 \\ \hline 112 \end{array}$$

112, area.

And the circumference is 38.

To make it more plain, I go through another operation, and suppose 6 to be the diameter. Now, 18 angles, radius 3 inches; take 6 and $\frac{1}{3}$ of $\frac{1}{6}$, equal to $\frac{1}{18}$, add to 6, equal 19, circumference. And of 18 angles and $\frac{1}{3}$ of $\frac{1}{18}$, equal 54, divide 54 by 2, equals 27; add 1, equals 28; multiply by 1 gives 28, area, and 19 inches, circumference.

I was told by an English gentleman, a lawyer, that there was no measure that was correct, for the want of a perfect quadrature of the circle; and experience has taught me that his words were true. By examining the table of long measure I find that in proportion as the circle is measured wrong, every other measure must and will be wrong; for the measure of the circle is what all measure springs from; it is the basis of all measures and weights.

Now, for instance, I have taken Archimedes' measure, as 7 to 22, which is the nearest to correct measure that I have ever found, and it is just $\frac{1}{4}$ of an inch in 33 too short, almost $\frac{3}{10}$ in a yard, and $\frac{1}{10}$ in a foot; and every other measure will be astray in the same proportion.

To find the area of a circle: On a subsequent consideration on finding the area of the circle by 3 times the square of the radius, I was thinking that sometimes the radius would be a hard number to square, as in a case where a surd number comes in; and as it is considered impossible to perfect the square of a surd number, I will give another way of finding the area, thus:

Multiply 3 times the radius, and the result will be the same as 3 times the square of the radius. Suppose 12 the diameter, 6 the radius; then 3 times 6 equal 18;

```
        18
         6
       ---
 27 ) 108 ( 4
      108
      ---
        0
```

Now add the 4 to 108, which gives 112, area.

To find the measure of a circle: 60 degrees is the only perfect triangle, and 6 is the only perfect number; so I take 6 of these triangles, and they form a hexagon, which is the basis of a circle, and there will be 6 sides and 6 radii. Then, to obtain the circumference, I square the radii by 6; $6 \times 6 = 36$, descending square. Then square the square:—

```
   36
   36
  ---
  216
 108
 ----
 1296
```

The biquadrate is equal to $\frac{1}{36}$ of $\frac{1}{36}$; and the sides of the hexagon being divided by 36, the two sides of the biquadrate will equal $\frac{2}{36}$ or $\frac{1}{18}$ of the side of the hexagon. So, as the biquadrate gains $\frac{1}{18}$ on one side of the hexagon, it will gain $\frac{1}{18}$ of the 6 sides; therefore to 36 add $\frac{1}{18}$, equal to 2, which gives 38 for circumference.

THE RATIO FOR A CIRCLE.

In regard to finding a correct ratio, I was told by a professor in Connecticut that I never could do it, for I might figure half way round the earth, and it would never come right. I was told about the same by a professor in London. His reason was, because it was a surd number, which I allow has been called an impossibility. This gentleman in London, to convince me that it is impossible, takes his pencil, and divides 19 by 6, thus:— 6) 19 (31666, and says it will always be 6, which proves it to be an impossibility, and throws away his pencil. I knew that this was one of the greatest difficulties — to find a correct ratio — and to do it I had to find a correct way to work out the surd number, which I have laid down in my work, as showing how I came by my ratio.

As I have before said, the measure of the circle is the only perfect and independent measure, from which all other measures spring, and without which no measure can be proved or worked right. We find in the table of long measure that 3 barleycorns make an inch. This is imperfect, as the length of a barleycorn is not known. As 7 to 22 is the nearest correct ratio that has come to my knowledge, and this is $\frac{1}{6}$ of an inch short of measure, the cubic inches in a yard would be 46656; but by the ratio 6 to 19 there would be 47724; so it lacks $\frac{1}{132}$ in length, equal to $553\frac{70}{130}$; and as there are 3 sides to a cube, I must multiply by 3:—

$$\begin{array}{r} 553\frac{70}{130} \\ 3 \\ \hline 1060\ 78 \\ 8, \text{ add for angle.} \\ \hline 1068, \text{ short of measure.} \end{array}$$

$$\begin{array}{r} 46656 \\ 1068 \\ \hline 47724.78, \text{ correct measure.} \end{array}$$

So the present measure falls short as follows: Linear measure, 75¾ per cent., square measure, 1.52½ per cent., cubic measure, 2.29 per cent.

To find the circumference of a circle 12 inches in diameter: Take the hexagon, composed of 6 triangles, having 18 angles. Now, one angle on the hexagon is 6 inches; so I take 2, equal to $\frac{1}{18}$ of the whole, and of that I make an oblong square, 6 inches long by 2 inches wide, with a line through the centre.

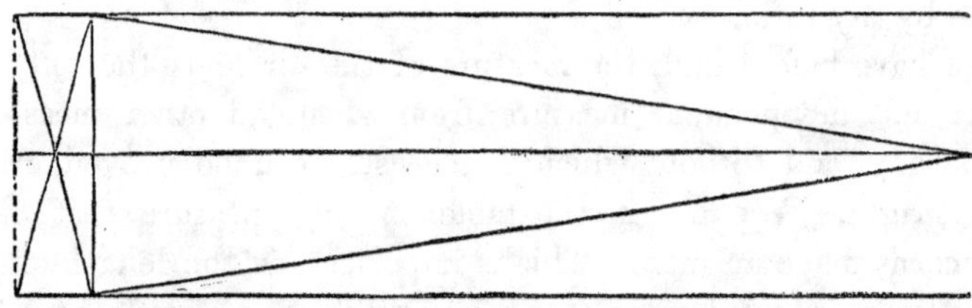

The two sides of the oblong square represent the two radii; and if a square be placed on the top, and drawn one side, and then drawn into the centre, it will form a cone, as is seen in the figure; and the lines gained on the top will strike the sides of the oblong just $\frac{1}{18}$ above the top of the cone, which shows that a line drawn across the top of them to the two sides of the cone would be just $\frac{1}{18}$ longer than $\frac{1}{18}$ of the hexagon, which would be just the measure of that part of the circle; so it proves that the circle gains $\frac{1}{18}$ of the hexagon; and if the hexagon equal 3 times the diameter, or 36, add $\frac{1}{18}$, equal 2, which gives 38, the circumference.

If the radius is 6 inches, then the sides of the square, by moving 1 inch to the centre point, in 6 inches will move just $\frac{1}{6}$ of an inch to the centre line, equal $\frac{1}{3}$ in the whole width. And 6 is equal to 18 thirds; now to 18 add 1, = 19; twice 18 are 36, and $\frac{1}{18}$, equal 2, added to 36, equals 38, the circumference.

To prove that the angle to measure a circle should be 60 degrees, instead of 90: I find, by examining books on this subject, that they make the area of a circle 12 inches in diameter 114 inches, allowing it to be 38 in circumference. I should make it the same if I wrought it by 90 degrees, as 3 times the square of the radius, $6 \times 6 = 36, \times 3 = 108$; add 6 to 108, = 114. But I find the measure by the biquadrate, and that is but 2 inches; the square of 2 is but 4; so $108 + 4 = 112$, instead of 114. So, if 90 give 6, 60 will give 4; add the 4 to 108, which gives 112.

Proof. — Take an angle of the hexagon, one half of which is 3, and this, when divided, makes the length of leg only $1\frac{1}{2}$;

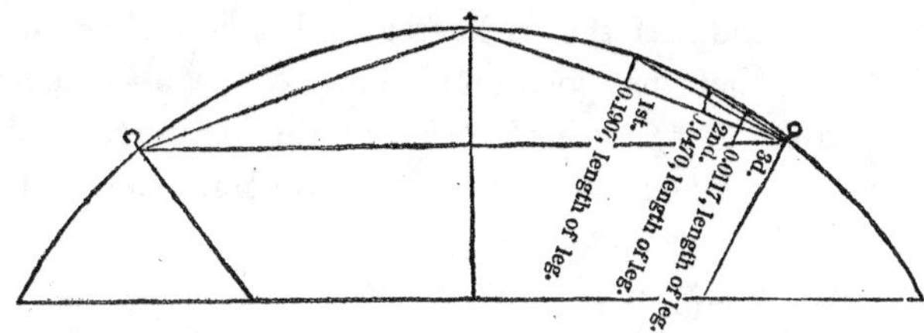

so I make it all halves; and as the radius is 6 inches, I multiply by 12 halves: —

```
 2494      40000           12
  400      39786           12
 ----      -----          ---
 2094     0.0214          144
   19                       9   Subtract 1½ = 3, × 3 = 9.
 ----                     ---
1.8846                    135 ( 11.6187
2.094                       1
------                    ---
3.9786                 21 ) 035
                             21
                            ---
                      226 ) 1400
                            1356
                            ----
                      2321 ) 4400          60000
                             2321          5.8093
                             ----          ------
                     23328 ) 203000        0.1907, 1st leg.
                             185824           470, 2d leg.
                             ------           117, 3d leg.
                    232367 ) 1807600       ------
                             1626369        .2494
```

But as the outside of the oblong square and the polygon is taken 5 times, and should be taken but once, it is $\frac{4}{100}$ too much, which must be subtracted. Then multiply the remainder by 19, equal one half the perimeter, and it falls short of 4 inches by $\frac{214}{10000}$; but the polygon $\frac{3}{8}$ long, &c., will make it good. So you see that 112 is the area of the perimeter; and if the area is right, when made into rings, as before shown, the measure of the circle cannot be wrong.

I have measured the circle, and find it gains just the biquadrate; but there are few that can find the perfect roots; if they could, they would have saved me much trouble, in respect to my problems; for if they were masters of the square root they might soon see that they could not get $113\frac{1}{10}$ inches within a circle of 12 inches diameter and $37\frac{7}{10}$ perimeter, when I can get only 112 in one of 12 inches diameter and 38 perimeter. See the diagram of rings on a previous page. The average measure of the rings is $3\frac{1}{9}$ inches, and they are 36 in number. Multiply 36 by $3\frac{1}{2} = 112$, area; and if this measure is correct, it is surprising that the measure of the circle should be wrong. Let common sense dictate; I ask no more.

Again: the groundwork of a circle is the hexagon, with 6 triangles and 18 equal sides. Now, the radius is $\frac{1}{3}$ of $\frac{1}{6}$, equal $\frac{1}{18}$; and as every thing in a circle has equal proportions, the curve must therefore be equal; and as the circle is the most perfect thing in nature, every part must have a natural and proportional result. The hexagon gives $\frac{1}{3}$ of $\frac{1}{6}$ for radius; consequently, the radius must give $\frac{1}{3}$ of $\frac{1}{6}$ for curve, in every part; and $\frac{1}{3}$ of $\frac{1}{6}$ is $\frac{1}{18}$ of itself; so the circle gains $\frac{1}{18}$ on the radius, and the radius is $\frac{1}{18}$ of the 6 angles of the hexagon, mark as you like; and it will be the same as $\frac{1}{3} \times \frac{1}{6} = \frac{1}{18}$; and $6 \times 3 = 18 + 1 = 19$, perimeter.

TAKING THE HYPOTENUSE.

As the proportion of 7 to 22, the most correct that has come under my observation, falls $\frac{1}{6}$ of an inch short of measure, I will show the number of cubic inches in a yard by this measure, and by the true ratio, as 6 to 19:—

36
36
——
216
108
——
$\frac{1}{3}$ of 132=44) 1296 (29.5 Add this 29.5 to 1296
88 29.5
—— ——
416 1325.5, the square.
396 36
—— ——
200 7953.0
39765
——
47718, from as 6 to 19.

```
       36
       36
     ----
      216
     108
     ----
     1296
       36          47718, from as 6 to 19.
     ----          46656, from as 7 to 22.
     7776          -----
    3888            1062, difference.
    -----
    46656, from as 7 to 22.
```

This is worked by taking the hypotenuse of a circle. Suppose it is 12 inches in diameter; first I take the radius, 6, for hypotenuse; one side of a polygon of 12 sides equals 3, and ½ of 3 is 1½, for known leg; and as it comes halves, I multiply by 2, to bring it all into halves, saying twice 6 is 12, and 3 × 3, = 9, subtracted:—

```
     12
     12
    ---
    144
      9
    ---
    135, as in the work.
```

The second operation is but ⅔ of an inch on the circle; therefore, multiply by 4:—

```
      6
      4
    ---
     24
     24
    ---
    576
      9
    ---
    567, as in the work is shown.
```

```
      12
      12      3
     ----     3
     144      -
       9      9, subtracted.
     ----
21 ) 135 ( 11.61895      2 ) 11.61895
      35                    --------
      21                     5.80945
     ----                   --------
226 ) 1400                   600.000, radius.
      1356                   580.945
      ----                  --------
 2321 ) 4400                  19055  These two added together
        2321                   1000         make just 6 inches,
        ----                  -----         and the oblong
23228 ) 207900                18055, 1st.   square 1/100 too
        185824                 3706, 2d.    much.
        ------                -----
232369 ) 22076               217610
         20913                  18 3/10
         -----                -------
2323785 )                     652830
                            1.740880
                            2.17610    1 7/10 increase of circle.
                            --------
                            3.999263 toward the biquadrate of
                                        a circle 12 inches in
The biquadrate is 4.000000              diameter.
                  3.989895 toward 4,0.
                  --------
                   .000737, the lack of 4 inches in the
                                 biquadrate.
```

I was called on to go to No. 356 Broadway, New York, to see what was called the squaring of the circle by two gentlemen from Cuba — Don Lewis de la Torrey Sages and Senor Sedano. They cut from a piece of metal — zinc or brass — a circle about 4 inches in diameter. They then cut this into 26 pieces, and placed them in the circle where the piece was taken from, which filled the circle to perfection. They then knocked the 26 pieces promiscuously on the table, and took them one at a time, and formed with the pieces a perfect square.

This is what I thought to be impossible. I think it was a most unexpected feat, and worthy of note and observation.

This I claim as one of the convincing proofs of my measure of the circle, so far as mechanical operation can go to demonstrate it.

My biquadrate, which is what makes up or measures the circle, is by them demonstrated to prove my measure. They cut the circular part, which is my biquadrate, to a square, which is evidence to the eye that my method is correct.

Any person can see that if the circular part of a circle 12 inches in diameter is cut off, so as to leave a perfect square, the difference in the circle obtained from as 7 to 22 and that obtained from as 6 to 19 will be clearly perceptible in the circular part; because the difference in a circle 12 inches in diameter, from as 7 to 22 or as 6 to 19, is $\frac{3}{10}$ of an inch, which I think is perceptible to any one who looks with an honest eye.

Notwithstanding all this, I claim that no mechanical demonstration can prove the perfect measure of the circle, because mathematics will discover that point which the most discerning eye cannot discover. The mathematical evidence which I have exhibited in my work far exceeds any mechanical operation.

THE ANGLES, ETC.

A circle is a plane figure, which has an equal extension in every direction from its centre to its circumference.

The diameter of a circle is a straight line passing through its centre, and extending in length to the extreme limits of the circle.

The radius of a circle is a straight line drawn from the centre to the circumference.

The radius of a sphere is a straight line drawn from the centre to the surface.

An angle is the opening between two lines which meet each other at a point.

One straight line is perpendicular to another when the angles on each side of the perpendicular are equal to each other.

Angles made by lines meeting each other perpendicularly, or crossing each other perpendicularly, are called right angles.

An acute angle is one that is smaller or sharper than a right angle.

An obtuse angle is one that is larger or more open than a right angle.

A triangle is a plain figure, enclosed by three straight lines of circumference.

An equilateral triangle is one whose three sides are all of equal length.

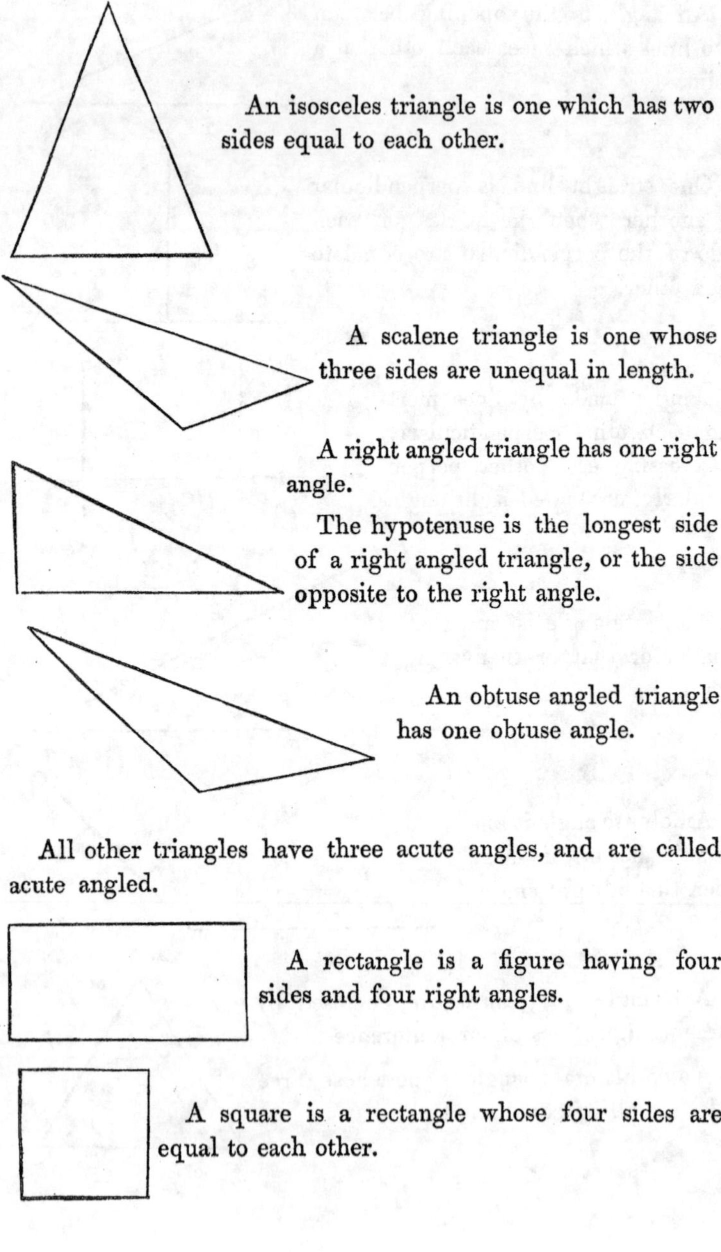

An isosceles triangle is one which has two sides equal to each other.

A scalene triangle is one whose three sides are unequal in length.

A right angled triangle has one right angle.

The hypotenuse is the longest side of a right angled triangle, or the side opposite to the right angle.

An obtuse angled triangle has one obtuse angle.

All other triangles have three acute angles, and are called acute angled.

A rectangle is a figure having four sides and four right angles.

A square is a rectangle whose four sides are equal to each other.

A rhombus is a plane figure, having four equal sides and two obtuse and two acute angles.

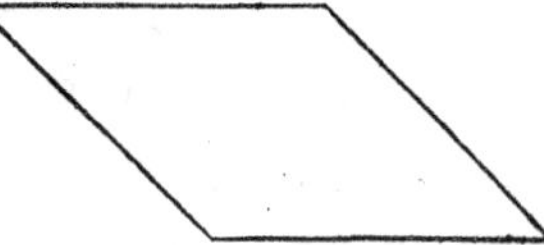

A parallelogram is a plain figure with four sides, having the opposite sides parallel to each other; and as parallel lines every where preserve an equal distance between them, the opposite sides of parallelograms are equal.

An arc of a circle is any part of its circumference.

The chord of an arc is a straight line joining the two extremities of the arc.

The segment of a circle is a part of the area cut off by a chord, or the part inclosed by an arc and its chord.

An axiom is a self-evident truth, or one so manifest that it cannot be made more clear by any demonstration.

The sign +, when placed after a number, denotes that more is to be added in order to complete the result.

This sign —, called minus, when placed after a number, denotes that something is to be taken from it.

This sign =, when between two numbers, denotes equality.

This sign × denotes multiplication.

This sign ÷, placed between two numbers or quantities, denotes that the one is to be divided by the other.

This sign $\sqrt{}$ denotes that the square root is to be extracted.

This sign $\sqrt{}^3$ denotes that the cube is to be extracted.

A polygon is the general name applied to plane figures with any number of sides.

A hexagon is a figure with six sides, equal in all its parts, different from all other figures, and will measure a circle, with the number 6.

MECHANICAL POWERS.

The mechanical powers are certain simple machines used for raising greater weights, or overcoming greater resistance, than the natural strength of man can perform without them.

These simple machines are six in number, viz.: 1, the lever, 2, the wheel, 3, the pulley, 4, the screw, 5, the wedge, 6, the inclined plane.

Force is a power exerted on a body, to move it; if it acts instantaneously it is called percussion; if it impels constantly it is an accelerative force.

Gravity is that force which causes a body to fall downward. It is called absolute gravity when in an empty space, and relative gravity when immersed in fluid.

Specific gravity is the proportion which the weight of one body bears to another.

The centre of gravity is a certain point in a body, upon which, when suspended, it will rest in any position.

The centre of motion is a fixed point round about which a body moves; and the axis of motion is that fixed line about which it moves.

Power and weight, when opposed to each other, signify one body that moves another; and the body that moved the body which communicates the motion is the power, and that which receives the motion is the weight.

Friction is the resistance which any machine suffers by the parts rubbing against each other. In the operation of machines, though all bodies are rough, in some degree, and all engines imperfect, yet it is necessary to consider all planes as perfectly even, all bodies perfectly smooth, and all bodies and machines to move without friction or resistance; all which sounds very plausible.

ON MECHANICAL POWERS.

The whole principle of relative motion in mechanics depends upon this one single rule: That the whole force of a moving body is the result of its quantity of matter multiplied by the velocity of its motion.

Thus, when the products arising from the multiplication of the particular quantities of matter in any two bodies by their respective velocities are equal, the entire forces are also equal. For example: suppose a body which weighs 40 pounds to move at the rate of 2 miles in a minute, and another body, that weighs 4 pounds, to move at the rate of 20 miles in a minute; the entire force with which these two bodies will strike against each other would be equal powers to stop them; for 40 multiplied by 2 gives 80, the force of the first body, and 80 is also the product of 4 multiplied by 20, the force of the second body. Therefore, the heavier any body is, the greater is the body or power required either to move or stop it; and the swifter it moves, the greater is the force. When two bodies suspended on any machine are put in motion, and the perpen dicular ascent of one body multiplied into its weight is equal to the perpendicular descent of the other body multiplied into its weight, those bodies, however so unequal in their weight, will balance one another in all situations; for as the whole ascent of one is performed in the same time as the whole descent of the other, their respective velocities must be directly as the spaces through which they move; and the excess of weight in one body is compensated by the excess of velocity in the other. Upon this principle the power of any machine may be easily computed; for it is only necessary to find how much swifter the power moves than the weight does, that is, how much faster in the same time, and just in that proportion so much power is gained by the engine.

The Lever is a bar, either of iron or wood, one part of which is supported by a prop at its centre of motion; and the velocity of every part or point in the lever is directly as its distance from the prop. There are four kinds of levers: 1st, the common lever, in which the prop is placed between the weight and power, but much nearer the weight than the power; 2d, when the prop is at one end, the power at the other, and the weight between; 3d, when the prop is at one end, the weight at the other, and the power is applied between them; the 4th differs from a lever of the first kind only in being bent. Levers of the first and second kind are often used in engines; the third kind is seldom used, as no power can be gained by them. When the power is at the same distance from the prop that the weight is, and the power and weight are both alike, the machine will remain in equilibrio, and no power can be gained. This is the principle upon which the common balance is formed.

ELECTRICITY.

The earth, air, and terrestrial bodies are supposed to contain a certain quantity of an elastic substance or subtile fluid, called by philosophers the electric fluid; and when any body possesses more or less of this fluid than what naturally belongs to it, several effects are visible in it, and the body is said to be electrified. This certain quantity of electric fluid found in all bodies could never be increased or diminished if all bodies admitted the passage of this electric fluid through their pores, or along their surface; but there are many bodies which will not suffer this fluid to pass through them, while others freely permit it. Those bodies through which the electric fluid can pass are called conductors of electricity, of which the most perfect are all kinds of metals. Those bodies through which the electric fluid cannot pass are called non-conductors of electricity; of which the most perfect are, glass, resin, sealing wax, sulphur, beeswax, and baked wood, among solids, and oils and air among fluids; but all substances become conductors when they are very hot. Conducting substances are also called non-electric, and non-conducting substances are called electric. Into these two classes all bodies are divided by electricians.

When any body has acquired an additional quantity of electric matter, and is surrounded with non-conductors, or bodies through which the electric fluid cannot pass, it must remain overloaded; and if it has lost a part of its natural share of electric matter it must remain exhausted, because the bodies which surround it prevent any of the electric fluid from entering into or coming out of it; and the body is then said to be insulated.

There are two principal theories of electricity, each of which has had its advocates. The one is, the two distinct electric fluids, repulsive with respect to themselves, and attractive of one another, adopted by Mr. Du Fay, on discovering the two opposite species of electricity, and since remodelled by Mr. Symmes

PNEUMATICS.

Pneumatics is that part of natural philosophy which treats of weight, pressure, and electricity of the air, with the effects arising from them.

The air is that thin, transparent fluid body which surrounds the whole earth to a considerable height, and which, together with the vapors and clouds that float on it, is called the atmosphere. That the air is a fluid is evident, from the following properties which it possesses, in common with all other fluids, viz.: 1, it yields to the least force impressed on it; 2, its parts are moved among one another; 3, it presses according to its perpendicular height; 4, and its pressure is every where equal.

The reason the earth turns upon its axis is, the weight of air, there being 15 pounds' weight to every square inch upon the earth's surface; that is, every square inch of earth sustains 15 pounds' weight of air. The moon has no atmosphere; if it had it would turn upon its axis, the same as our earth, by its atmosphere, or weight of air. The moon does not attract but repels the water, which causes the ebbing and flowing of tides.

If the moon attracted, like the sun, the water would be quiet and still at the full moon, instead of ebbing and flowing. But the moon repels and the sun attracts, which causes the ebbing and flowing of the tides.

"By what way is the light parted which scatters the east winds upon the earth?" — *Job* 38: 24. Answer: In the first place, the weight of air on every square inch is calculated to be near 15 pounds. Now, I leave the college calculations, and take my experience as a mechanic. Suppose you are working machinery; by a good regulator you will find that your machinery takes more measure of water after sunrise than before. The reason of which is, the weight of the element is less on the surface, and it being more dense to the west, in the dark, than to the east, in the light, it rushes towards the east at the rate of more than 1000 miles per hour, and carries the earth round with it, which is what makes the earth's diurnal motion, and scatters the winds upon the earth. I think the wind blows more from the west than any other direction; but if the above was not the case, it would be like to blow from the east more than 1000 miles per hour.

This question has never before been answered, I think. When it seems to us a calm, the wind blows more than 1000 miles per hour.

CHANGE OF THE AIR.

The following statement shows that the rarefaction of the air at a distance from the earth's surface increases in a geometrical proportion, while its height from the earth increases in an arithmetical proportion: 7 miles above the surface, the air is 4 times lighter; 14 miles, it is 16 times lighter; 21 miles, 64 times lighter; 28 miles, it is 250 times lighter; 35 miles, it is 1024 times lighter; 42 miles, it is 4096 times lighter; and so on, in this proportion. From this it may be proved that a cubic inch of such air as we breathe on the surface would be so much more rarefied at the height of 500 miles that it would fill a sphere equal in diameter to the orbit of Saturn.

The weight or pressure of air is determined thus: fill with purified quicksilver a glass tube about three feet long, and open at one end, and putting your finger upon the open end, turn that end downward, and immerse it in a small vessel of quicksilver, without letting in any air; then take away your finger, and the quicksilver will remain suspended in the tube about $29\frac{1}{2}$ inches above the surface of that in the vessel. Therefore the air's pressure on the surface of the earth is equal to the weight of $29\frac{1}{4}$ inches' depth of quicksilver all over the earth's surface, at a mean rate. A square column of quicksilver $29\frac{1}{2}$ inches high and 1 inch thick weighs just 15 pounds, which, therefore, is equal to the weight of air upon every square inch on the earth's surface; and the weight upon a square foot, 144 inches, amounts to 2160 pounds. According to this, a middling-sized man, whose surface is generally about 14 square feet, sustains a pressure of air of 30,240 pounds, when the air is of mean gravity.

As the earth's surface contains near 200,000,000 square miles, in round numbers, and every square mile 27,874,400 square feet, there are 5,575,680,000,000,000 square feet on the earth's surface; which multiplied by 2160 pounds, the weight on each square foot, gives 12,043,468,800,000,000,000 pounds for the pressure of the whole atmosphere.

All common air is impregnated with what is called the vivifying spirit, which is essential to preserve life; and in a gallon of air there is enough of it for one man during the space of one minute, but not much longer. This spirit is also in the air which is contained in water, as appears by fishes dying when they are excluded from fresh air, as in a pond that is frozen over.

This spirit in air is lost by passing through the lungs of any animal, which is the reason why an animal dies so soon when deprived of fresh air. The little eggs of insects, also, when stopped up in glass, and excluded from the air, do not produce their young, though they be assisted by warmth. The seeds of plants, though mixed in good earth, will not grow if deprived of air.

The vivifying qualities are also destroyed by the air passing through fire, particularly charcoal fire, or the flames of sulphur. Air may also become vitiated by being closely confined in any place for a considerable time, or by being mixed with malignant vapors; and lastly, by being corrupted by vitiating spirit, as in the holds of ships, or in oil cisterns and wine cellars which have been shut for some time, or in brewers' vaults. In any of them the air may be so much vitiated as to cause immediate death to any animal that enters.

When the air has lost its vivifying spirit it is called damp, because it abounds with humid and moist vapors, and destroys life. This is known by those who work in mines.

When a part of the vivifying spirit of the air in any country begins to putrefy, the inhabitants of that country will be subject to an epidemical disease, which will rage till the putrefaction is over. And as the putrefying spirit occasions the disease, so if the diseased bodies contribute towards the putrefaction of the air, the disease will become pestilential and contagious.

THEORY OF THE WINDS.

Wind is the consequence of the rarefaction of air, and is no other than air put into motion by heat, or any other cause; for when the air is rarefied by heat it will swell, and thereby affect the adjacent air; and the degrees of heat being different in different places, there will arise various winds. When the air is heated to any degree it will ascend upward, and the adjacent air will rush in to supply its place; therefore there will be a stream or current of air from all the adjacent parts towards the place where the heat is. This appears evident from the motion with which the air rushes towards any place where there is a great fire, as in a glass house, or through the key hole of a door in a room where there is a fire.

That wind called the trade wind, which blows constantly from the east to the west, about the equator, is a necessary consequence of this principle; for when the sun shines perpendicularly upon any part of the globe, the air in that part will be heated, and, consequently, rarefied, and will therefore rise up. When the sun withdraws, the adjacent air rushes in to fill the place of the rarefied air, which will cause a stream or current of air from all parts towards that part which is most heated by the sun; and the course of the sun being from east to west, with respect to the earth, the common course of the air which supplies the place of the rarefied air must be in the same direction, viz., from east to west; but on the north side its course will be directed a little towards the north, and on the south side as much towards the south.

This would be the general cause of the wind about the equator, if it were not affected by other causes which change its direction, as, 1st, by exhalations that arise out of the earth at different times and different places, occasioned by subterraneous fires, volcanoes, &c.; 2d, by a sudden inundation of rain, which causes an extraordinary rarefaction of the contiguous air; 3d, by high mountains, which change the course of the winds; 4th, by the declination of the sun towards the north or south, thereby causing a greater heat in the air on the same side of the equator.

The following are the principal causes which create such a great variety and uncertainty of the winds in most countries distant from the equator: 1, the variations of the winds in different parts of Europe; 2, the monsoons, which are found in the Indian seas; 3, those winds which always blow from west to east on the western coasts of America, and on the east of Guiana, and the sea breeze which in hot countries blows from sea to land in the day time, and the land breezes which blow towards the sea in the night time, and all others.

THE CAUSE OF THUNDER AND LIGHTNING.

The effluvia and vapors arising from different bodies meet and unite together in the atmosphere, which is the common receptacle of all vaporous bodies, as the steam from most bodies, the smoke from bodies burned, and the effluvia emitted from sulphurous, nitrous, acid, and alkaline substances, and every volatile body, rises to a certain height in the atmosphere, according to its own specific gravity; and when the effluvia which arise from an acid and alkaline body meet each other in the air, there will be a conflict between these two vapors, or what is commonly called fermentation, between them. If this fermentation be great, it will produce fire; and if the effluvia be of a combustible nature the fire will run from one part of the air to another, following the inflammable matter.

These things may be demonstrated by the following experiment: Mix some oil of cloves and Glauber's spirits of nitre together, which will immediately produce a sudden fermentation, with a fine flame; and if the ingredients be wet, there will be a sudden explosion. These are the effects of the union of an acid and alkaline fluid.

From this experiment we may account for the effects of thunder and lightning, which are occasioned by the effluvia of sulphurous and nitrous bodies meeting each other in the air, where, assisted by the sun's heat, a fermentation, fire, and ex plosion ensue. When the inflammable matter is thin and light it will ascend to the upper parts of the atmosphere before the fermentation, fire, and explosion take place; but when it is more dense it will hover round the surface of the earth, where, when an explosion takes place, the fire is visible, and often dangerous; the explosion also has a violent force, and the heat, being great, will rarefy and drive away all adjacent air, kill men and cattle, split rocks and trees, &c. Lightning differs from all other fire; it has been known to pass through leather, wood, cloth, and other substances, without heating them; and at the same time melting iron, steel, silver, gold, and hard metals and bodies. It has melted or burned asunder a sword, without hurting the scabbard, and melted money in a man's pocket, without hurting him or his clothes. So fine are the particles of this fire that they pass through soft, loose bodies, without injuring them, and spend their force upon those more dense. Any steel instruments, as knives and forks, &c., which have been struck with lightning, have a strong magnetic virtue, which they retain for many years. The lightning has often turned the magnetic needle round, and made it point to the south pole, instead of to the north.

The explosions which sometimes happen in mines, and are called fire damps, are of the same nature with lightning, and are occasioned by sulphurous and nitrous vapors rising from the mine, which, mixing with the air, take fire from the light used in the mine. This fire, when once kindled, continues to run from one part of the mine to another, wherever the combustible happens to be; and as the electricity of the air is increased by the heat, the air in the mine will expand, and, for want of room, will explode, with a degree of force equal to the violence of the fire, the quantity of effluvia, and density of the vapors. This is sometimes so strong as to blow up the mine; at other times so weak that when it has taken fire it may be easily blown out. Air that will take fire from the flame of a candle may be produced thus: Having pumped the air out of the receiver of an air pump, let the air run into it through the flame of the oil of turpentine; then move the cover of the receiver, and holding a candle to the air, it will take fire.

When combustible vapors are kindled in the bowels of the earth, where there is little or no vent, they produce earthquakes, and as soon as they break forth into the open air, violent storms, or hurricanes of wind. An artificial earthquake may be produced thus: Take 10 or 15 pounds of sulphur and an equal quantity of the filings of iron, and knead them with common water into the consistency of paste; this being burned under ground will in 8 or 10 hours' time burst into flames, and cause the earth to tremble around it to a great distance. It is owing to substances of this kind and nature that we have volcanoes.

HYDROSTATICS.

Hydrostatics treats of the equilibrium of fluids, or the gravitation of fluid bodies remaining at rest. When this equilibrium is removed, and the fluid body set in motion, the effects that it then produces are called hydraulics.

The siphon is a bent tube.

A valve is a kind of flap or cover fixed to a pipe, or to the aperture of any body, and which, by opening only one way, suffers water or any fluid body to pass, but not to return.

A piston is a small cylinder fixed to the end of a rod, and fitted to the bore of a pipe, and frequently contains a valve.

AXIOMS.

All fluids except air are incompressible, or incapable of being compressed into less space or shape.

In a vessel of water, or any other fluid body, the pressure of the upper parts on the lower is in proportion to the depth, and is the same at the same depth, whatever the diameter of the vessel may be.

The pressure of a fluid upward is equal to the pressure downward at any given depth.

The bottom and sides of a vessel are pressed by the fluid it contains in proportion to its height, without any regard to the quantity.

If fluids of different gravities be contained in the same vessel, the heaviest will be at the bottom, the lightest at the top, and the others farther distant from the top, in proportion to their respective gravities.

The direction of the pressure of a fluid against the sides of the vessel that contains it is in lines perpendicular to the sides of such vessel.

A body that is heavier than an equal quantity of fluid will sink in that fluid; but if lighter it will swim at the top of the fluid; and if it be of the same gravity it will neither sink nor swim, but will remain suspended in any part of the fluid.

A solid immersed in a fluid is pressed on all sides by the fluid in proportion to the height of the fluid above the solid; and bodies very deeply immersed in any fluid may be considered as equally pressed on all sides.

Every solid immersed in a fluid that is specifically lighter loses as much of its own weight as is equal to the weight of a quantity of that fluid of the same dimension with the solid.

The fluid in which the solid is immersed acquires the weight the solid loses.

As the principal fluid with which we have any concern in hydrostatics is water, it may be necessary to name a few of its distinguishing qualities or properties: —

1. Water is a transparent, colorless, scentless fluid, which, with a certain degree of cold, turns to ice.

2. Water is one of the constituent parts of all bodies, as has been proved by distillation; for the earth, bones, the driest wood, and stones pulverized constantly yield a certain quantity of water.

3. Though fluidity is commonly regarded as an essential property of water, yet many philosophers, particularly Dr. Black, of Edinburgh, consider it as an adventitious circumstance, and produced by a certain degree of heat; and they assert its natural state to be crystalline, as when frozen.

4. Water is a more penetrating body than air, though it be less transparent; for it will pervade bodies that air will not, as is evident from its passing through the pores of a bladder.

5. Some bodies are dissolved by water, as salts, while it conglutinates others, as brick, stone, and bones.

6. Water owes its fluidity to heat, and it contains no small quantity of air; and the sediment found in all water which has not been distilled always contains a quantity of earth, from which last element it is supposed that plants derive all their nourishment.

A TABLE OF SPECIFIC GRAVITY.

1. — *Solids.*

Substance	Specific gravity	Substance	Specific gravity
Platina,	23000	Crystal,	2210
Fine gold,	19640	Clay,	2160
Standard gold,	18888	Oyster shells,	2092
Lead,	11325	Onyx stone,	2092
Fine silver,	11091	Brick,	2000
Standard silver,	10535	Common earth,	1984
Copper,	9000	Nitre,	1900
Copper halfpence,	8915	Vitriol,	1880
Gun metal,	8784	Alabaster,	1874
Fine brass,	8350	Horn,	1840
Cast brass,	8000	Ivory,	1825
Steel,	7850	Sulphur,	1810
Iron,	7645	Chalk,	1793
Pewter,	7471	Solid gunpowder,	1745
Cast iron,	7425	Alum,	1714
Tin,	7320	Dry bone,	1600
Lapis calaminaris,	2000	Sand,	1520
Loadstone,	5000	Lignumvitæ,	1327
Mean specific gravity of		Coal,	1250
the whole earth,	4500	Jet,	1238
Crude antimony,	4000	Ebony,	1177
Diamond,	3517	Pitch,	1150
Granite,	3530	Rosin,	1100
White lead,	3160	Mahogany,	1063
Island crystal,	2720	Amber,	1040
Marble,	2705	Brazil wood,	1031
Pebble stone,	2700	Boxwood,	1030
Rock crystal,	2650	Beeswax,	955
Pearl,	2630	Butter,	940
Green glass,	2600	Oak,	925
Flint,	2570	Gunpowder, shaken,	922
Common stone,	2500	Logwood,	913

TABLE OF SPECIFIC GRAVITY, CONTINUED.

Ice,	908	Fir,	550
Ash,	800	Sassafras wood,	482
Maple,	755	Cork,	240
Beech,	700	New fallen snow,	86
Elm,	600		

2.—*Fluids.*

Quicksilver,	13600	Ale,	1028
Oil of vitriol,	1700	Vinegar,	1026
Oil of tartar,	1550	Tar,	1015
Honey,	1450	Common water,	1000
Spirits of nitre,	1315	Distilled water,	993
Aqua fortis,	1300	Red wine,	990
Treacle,	1290	Proof spirits,	931
Aqua regia,	1234	Olive oil,	913
Human blood,	1054	Pure spirits of wine,	866
Urine,	1032	Oil of turpentine,	800
Cow's milk,	1031	Ether,	726
Sea water,	1030	Common air,	1.232, or $1\frac{7}{56}$

As these numbers are the weight of a cubic foot, or 1728 cubic inches, of each of the foregoing bodies, in avoirdupois ounces, the quantity in any other weight, or the weight of any other quantity, may be found by proportion. For example: required the contents of an irregular block of common stone, weighing 100 pounds, or 1792 ounces. Here, as 2500, the ounces in a cubic foot of common stone, is to 1792, so is 1728, the inches in a cubic foot, to $1238\frac{4}{5}$ cubic inches, the contents.

THE DIVISIONS AND SOUNDS OF LETTERS.

In the early ages of antiquity, before alphabets were invented, mankind, sensible of the want of some means of recording events, and historical and scientific discoveries, had recourse to various arts for these purposes, the first of which was painting

That partiality for pictures, so evident in all ages and countries, afforded the ancients a method of perpetuating their actions.

To commemorate that one man had killed another, they painted the figure of a dead man, with another man standing over him, with a hostile weapon in his hand.

On the first discovery of America, this was the only kind of writing used by the Mexicans.

The first improvement by our ancestors in the art of writing was by the introduction of hieroglyphical characters. These consisted of certain symbols, which were made to represent certain invisible objects. An eye was the symbol of knowledge; a circle was eternity, as having neither beginning nor end. The figures of animals were much employed in this kind of writing, on account of some quality of which they were supposed to be endowed, and in which they resembled the object signified. Thus imprudence was represented by a fly, wisdom by an ant, victory by a hawk.

These hieroglyphics flourished most in ancient Egypt, as did all other learning at that time, when the knowledge of these characters was reduced into a regular art; and they still exist in Egypt, in some degree.

The Chinese still use characters of this nature. They have no alphabet of letters, but every single mark or character signifies one perfect idea or object. The number of these characters is seventy thousand; and to make acquaintance with them would constitute a labor for the whole life of one man.

The next improvement in the art of writing was by the invention of signs or marks which stood not directly for the objects themselves, but for the words or names whereby they were distinguished. This was an alphabet of syllables. An alphabet of this kind is still in use in Ethiopia and some countries in India.

It has been considered that the discovery of the earth's traverse round the sun was new; but I find by Hodson that these systems have now given place to that called the Copernican system, which, undoubtedly, is the most ancient in the world

It was first introduced into Greece and Italy by Pythagoras, and from him called the Pythagorean system. It was adopted by Philolaus, Plato, Archimedes, and all the most ancient philosophers, but was at length lost under the Peripatetic philosophy, and restored again about the year 1500 by Nic. Copernicus.

This system has been proved, by the most evident demonstrations, to be the only true one. First is a representation of this system, where the seven concentric circles marked Mercury, Venus, the Earth, Mars, Jupiter, Saturn, and Georgium Sidus, represent the orbits of these seven primary planets, each performing its annual rotation round the sun, which is placed in the centre.

The next two circles represent the twelve signs of the zodiac, with all its divisions into thirty degrees in each sign; and, lastly, the next outer circles show the twelve calendar months of the year, with their divisions into days, each in its order.

TO CORRECT MEASURE.

As the foot is divided into 12 inches, so each inch is divided into 12 parts, called seconds, and each second is again divided into 12 thirds, and each third into 12 fourths, &c.

Now, according to correct measure, derived from a perfect quadrature of the circle, being in proportion of 6 to 19, instead of 7 to 22, which makes a difference of as $\frac{1}{132}$ is to $\frac{1}{133}$, the foot rule, which is called 12 inches, is near $\frac{1}{10}$ of an inch too short; so it will have to be made near $\frac{1}{10}$ of an inch longer, and the $\frac{1}{10}$ or near $\frac{1}{10}$ will be divided equally to each of the 12 inches.

By reason of the imperfection in the measure of the circle, every other measure, as well as all weights, are wrong in the same proportion. To rectify them, I take a stick or rod 6 feet long, and say, if 7 diameter gives 22 circumference, as 6 to 19 gives $22\frac{1}{6}$, so 3 times 22 are 5 feet 6 inches; add 3 times $\frac{1}{6}$, equal to 5 feet $6\frac{1}{2}$. The present measure, at 7 to 22, makes it on the rod 5 feet 6 inches; and divide 5 feet $6\frac{1}{2}$ inches into

11 equal parts, equal to $5\frac{1}{2}$ feet, which will be the correct measure in proportion as 6 to 19, from which you may take feet, inches, yards, &c., &c. And if the present measuring instruments are right, in proportion as 7 to 22, all the former tables will be right.

With respect to weights, the gravity of a cubic foot of water is $62\frac{1}{2}$ pounds, in proportion of as 7 to 22; but in proportion of as 6 to 19 it is 2.29 per cent. short; so the weights would require to be altered in the same proportion as the measures, and all would be right. The weights are full 36 hundredths of an ounce in a pound short weight, by the proportion of as 6 to 19.

MEASURES, ETC.

Mensuration in general is the art of measuring and estimating the magnitudes and dimensions of bodies or figures, and is divided principally into three parts, called linear measure, superficial measure, and solid measure.

1. Linear measure is measuring length, without breadth or depth.

2. Superficial measure consists of length and breadth taken together.

3. Solid measure consists of length, breadth, and depth.

4. A point has no parts nor diameter, neither length nor breadth.

5. A line has only length, without any other dimension.

6. A right line lies all in the same direction, and is the shortest way between its two extremities.

7. A curve line continually changes its direction.

8. Parallel lines are always at the same distance, and meet in no angle.

9. Oblique right lines change their distance, and meet in an angle.

10. An angle is the meeting of two lines.

11. If the two lines which form an angle be perpendicular to each other, they form a right angle.

12. But if two lines be not perpendicular to each other, they form what is called an oblique angle, which is either greater or less than a right angle.

13. An angle that is smaller than a right angle is called an acute angle.

14. An angle that is greater than a right angle is called an obtuse angle.

15. A triangle is a figure of three lines, and has various names, according to its angle.

16. An equilateral triangle has its three sides, and consequently its three angles, equal to each other.

17. An isosceles triangle has only two sides equal.

18. A scalene triangle has its three sides and angles unequal to each other.

19. A right angled triangle has one right angle.

20. An obtuse angled triangle has one obtuse angle.

21. An acute angled triangle has all its angles acute.

22. A figure of four sides is called a quadrangle, or a quadrilateral figure, and is either a parallelogram, a square, a rhomboid, a rhombus, a trapezium, or a trapezoid.

23. A square is an equilateral rectangle, having all its sides equal, and all its angles right angles.

24. A rhomboid is an oblique angled parallelogram.

25. A rhombus is an equilateral figure having all its sides equal, but its angles oblique.

26. A trapezium is a quadrilateral figure, but its opposite sides are not parallel.

27. A trapezoid has only two opposite sides parallel.

28. Plane figures having more than four sides are generally called polygons; but they receive particular names, according to their number of sides. Thus a polygon of five sides is called a pentagon; one of six sides a hexagon; one of nine sides a nonagon; one of ten sides a decagon; one of eleven sides an endecagon; one of twelve sides a dodecagon.

29. A circle is a plane figure, bounded by one circular line called the circumference, which is every where equally distant from the centre.

30. The radius of a circle is a right line drawn from the centre to the circumference.

31. The diameter of a circle is a right line drawn through the centre, and bounded at each end by the circumference.

32. An arc of a circle is any part of the circumference.

33. A chord is a right line joining the two extremities of an arc.

34. The segment of a circle is any part of it.

35. A semicircle is half a circle.

36. A sector is a part of a circle contained under part of the arc and two radii drawn to the centre.

37. A quadrant is a sector of a circle having one quarter of the circumference for its arc, and its two radii perpendicular to each other.

38. The circumference of every circle, in geometry, is supposed to be divided into 360 equal parts, called degrees, and each degree subdivided into 60 minutes, and each minute into 60 seconds, and so on. Hence a semicircle contains 180 degrees, and a quadrant 90 degrees, which form a right angle; and half a quadrant, called an octant, contains 45 degrees; for the measure of every right line angle is an arc of a circle contained between the two lines which form the angle, the point of the angle being in the centre of the circle; and the number

of degrees contained in the arc of the circle gives the measure of the angle.

39. In every right angled triangle the side opposite the right angle is called the hypotenuse, and the other two sides the legs of the triangle.

40. The height or altitude of a figure is a line drawn from the uppermost side or angle, perpendicular to the base.

41. An angle is generally described by three letters, the name of a polygon. Thus, one of three sides is a trigon; one of four sides a tetragon; one of five sides a pentagon; one of six sides a hexagon; one of seven sides a heptagon; one of eight sides an octagon; one of nine sides a nonagon; one of ten sides a decagon; one of eleven sides an endecagon; one of twelves sides a dodecagon.

TO MEASURE A SPHERE OR GLOBE.

Having given the measure of a circle, I will give a simple rule to measure a globe or sphere, having seen many errors in this measure, as no man could measure a globe without having the perfect measure of the circle.

To find the true measure of the surface of a globe, in feet, inches, &c.: Suppose 12 is the diameter; then the circumference will be 38 inches; so multiply half the circumference by itself, and the product will be the true measure of its surface, thus: —

$$\begin{array}{r} 19 \\ 19 \\ \hline 171 \\ 19 \\ \hline 361 \end{array}$$

This is the true measure of the surface of a globe 12 inches in diameter.

THE PYRAMID.

A pyramid is a figure that has a right lined figure for its base, and each of its sides is a triangle, whose vertices meet at a point at the top, which is called the vertex of the pyramid.

The pyramid takes its name from the figure of its base, like the prism. So a pyramid is a square stick of timber, at a given square at the base, and running to a point at the top.

To measure a pyramid: Square the diameter at the base; take half that product; multiply that by the height, which gives the square feet.

Suppose a pyramid be 100 feet high, and 4 feet square at the base, and run to a point; how many square feet are there in it?

Square the base, $4 \times 4 = 16 \div 2$.

```
                              4
                             ---
                           2) 16
                             ---
                              8
Multiply by the height,     100
                            ---
                            800 feet in the pyramid.
```

It has been the custom to take the mean diameter of the base, which, if the base be 4, will be 2; and $2 \times 2 = 4$; and as it is 100 feet high, $100 \times 4 = 400$ feet, solid contents. But my way is thus: 4 feet square, base; $4 \times 4 = 16$; then multiply by the height, 100, $= 1600$; divide the 1600 by 3, which gives $533\frac{1}{3}$ cubic feet.

THE PRISM.

A prism is a solid body, wood, stone, &c., as a stick of timber, square at each end, and its sides alike.

To find the measure of a prism: Multiply or square one end, and then multiply that product by the length, which will give the measure in cubic or solid feet.

Suppose a square stick of timber to be 100 feet long, and 4 feet square; how many cubic feet will it contain?

```
     4
     4
    ——
    16
   100
  ————
  1600 cubic feet in the above stick.
```

THE CYLINDER.

The cylinder is a round prism, being all its length circular.

To measure a cylinder: Suppose a cylinder be 100 feet long and 1 foot in diameter; how many square feet will it contain? Find the area of one end, and multiply that by the length, which will give the answer.

```
          112
           12
        ——————
1728 ) 134400 ( 77 feet, 9 inches, and 48/1728.
        12096
        ——————
         13440
         12096
         —————
          13440
          1296
          —————
           48, remainder.
```

Or you may measure a cylinder in this way: Suppose your cylinder to be square, and as large as the diameter; measure it the same as a square, and take $\frac{7}{9}$ of the product, which will give the circular measure. Thus, if your cylinder be 100 feet long and 1 foot in diameter: —

100
7

9) 700 (77 feet, 9 inches.
63

70
63

7

To measure a cone: Find the square of the base; then multiply by the height, and divide that product by 3, which gives the square contents; then take $\frac{7}{9}$ of the product, which gives the circular measure, thus: —

4
4

16
100

3) 1600

$533\frac{1}{3}$
7

9) $3733\frac{1}{3}$

$414\frac{7}{9}\ \frac{1}{27}$.

Suppose I had a stick of round timber to sell by the foot; the stick is 18 inches at butt, and 6 inches at the crop. I charge you one shilling a foot; you buy half the length to-day, and the other half to-morrow; how much did you lose by buying it at two purchases?

Now, according to the adopted measure of timber of this description now in use, you lost ninepence. The operation of this sum, mathematically, will demonstrate the measure of the circle, and show its use.

Suppose a round stick or block of timber 1 foot long and 17

inches in circumference had to be squared; how many inches would you lose in squaring? Answer, 80⅞ inches.

Multiply 2 and 6 pence by 2 and 6 pence.
Answer, 6 and 3 pence.

I have two half crowns, each equal to 5 shillings; I want them multiplied so as to make 6 and 3 pence.

What does £1 19 s. 11 d. 3 qr. multiplied into itself amount to? Answer, £3 19 s. 11 d. 3 qr.

Subtract 19 rods, 5 yards, 1 foot, 5 inches from 20 rods.
Answer, 1 inch.

Suppose a ball be shot from a gun, on a level, 3950 miles, and then fall to the ground, perpendicularly; what distance must I travel to find the ball, and how far would it have to fall from where it lost its force?

Answer: I should have to travel 3126¾ miles, and the ball would fall $1636\frac{1}{10}$ miles, and a little over.

Suppose I start from a centre, and travel on a radius 40 rods; what length of chain, each link to be 1 inch long, or how many links would it require for the circumference?
Answer: It would require 5016 links.

Suppose the reckoning of a party in a public house amounts to 6 shillings and 1 farthing; what number of persons must there be, to pay an equal share?
Answer: 17 persons, and each to pay 4¼ pence.

If ⅓ of 6 be 3, what will ¼ of 20 be? Answer: 7½.

What rule is that in multiplication which does not increase by multiplying? Answer: Cross multiplication.

Suppose a measure to be made in circular form, and to contain one bushel, allowing 2256 inches to the bushel; what is the diameter and depth?

Suppose a circle to be 17 inches in diameter; what is the circumference and area?

Answer: The circumference is 53 inches, and the area 224.77⅕ inches.

What is the number of square feet or inches in a cylinder 22 feet 6 inches long, and 4 feet 3 inches in diameter?

According to the measure now in use, from as 7 to 22, it would bring the world to an end in 66 years, which would be 132 revolutions, thus:—

$$\begin{array}{r} 22 \\ 6 \\ \hline 2\,)\,132 \text{ revolutions.} \\ \hline 66 \text{ years.} \end{array}$$

As short measure forms an internal scroll, and over measure forms an external scroll.

I have 36 straight strips, each 19 inches long and $\frac{1}{6}$ of an inch wide. I wish to know how many square inches there are in each strip, and how many square inches there are in the 36 strips, in proportion of square measure, and also in proportion of circular measure. When you work this, and understand it, you will see the difference between square measure and circular measure.

Suppose a box be 2 feet long; how wide and how high must it be to contain half a yard square?

Find the square of the length.

$$\begin{array}{r} 18 \\ 18 \\ \hline 144 \\ 18 \\ \hline 3\,)\,324 \\ \hline 108 \\ \hline \surd\ 216\,(\,14\tfrac{20}{24} \\ 1 \\ \hline 24\,)\,016 \\ 96 \\ \hline 20 \end{array}$$

The box is $14\frac{20}{24}$ by 24 long, $14\frac{20}{24}$ wide, and $14\frac{20}{24}$ high. The above is not perfect. I work it thus: —

```
                           18
                           18
                          ----
                          144
                          18
                          ----
                          324
                            18, to cube.
                         -----
                         2592
                         324
                         -----
24 inches long.    24 ) 5832 ( 243 square inches.
                         48
                         ----
                         103
                          96
                          ----
                           72
                           72
                           --
                            0

                  √ 243 ( 15.583, height and depth.
                    1
                    ---
               25 ) 143
                    125
                    -----
              305 ) 1800
                    1525
                    ------
             3108 ) 27500
                     24864
                     -----
                      2636
```

Suppose I have a round stick, 50 feet long and 18 inches at the base, and running to a point. I wish to know the length of a string that is wound 3 times round each and every foot, commencing at the base and running to the extreme of the 12 inches, the same as a barber's pole is painted.

I find the circle of the mean circumference, which is 9 inches: —

```
                9
                9.5
              -----
               81
                4½
              -----
           3 ) 85.5 ( 28.5, the circle.
Now I square the circle.   28.5
                28.5
              -----
               1425
              2280
              570
              -----
              812.25
```

This is what the string rises, 4 × 4 = 16 ;

```
              812.25
                16
              ------
            √ 828.25 ( 28.77933
              4
              ------
         48 ) 428
              384
              ------
        567 ) 4425
              3969
              ------
       5747 ) 45600
              40229
              ------
      57549 ) 537100
              517941
              ------
     575583 ) 1915900
              1726749
              -------
    5755863 ) 18915100
              17267589
              --------
               1647511
```

```
        28.77933
             150
        ---------
        143896650
        2877933
        ---------
   12 ) 431689950 ( 359 7.4162
        36
        --
         71
         60
         ---
         116
         108
         ---
           88
           84
           --
            49
            48
            --
             19
             12
             --
              75
              72
              --
               30
               24
               --
                6
```

Answer: The string will be 359 feet, 7.4162 inches.

A and B bought 300 acres of land for 600 dollars; each paid 300 dollars; A says to B, "Let me have my choice in the land, and my land shall cost me 75 cents an acre more than yours."

The sum is 65 to 95.

```
 80 : 150 : : 90                          80 : 150 : : 65
        80                                       80
     -----                                    -----
95 ) 12000 ( 126, answer.             65 ) 12000 ( 184, answer.
      95                                    65
     ---                                    ---
     250                                    550
     190                                    520
     ---                                    ---
      600             126, ans.              300
      570             184, ans.              260
      ---             ---                    ---
       30, remainder. 310                     40, remainder.
```

310 : 126 : : 300	310 : 184 : : 300
300	300
310) 37800 (121, ans.	310) 55200 (178, ans.
310	310
680	2420
620	2170
600	2500
310	2480
290, remainder.	20, remainder.

Answer: A has $121\frac{290}{310}$; B has $178\frac{20}{310}$.

To extract the circle or globe, after the manner of the square or cube root: Take any number of figures, say 448; divide the given number by 7, and add twice the quotient to the given number; extract the square root of the same, and it will be the diameter.

7) 448 (64	64	448
42	2	128
28	128	576 (24, diameter.
28		4
0		44) 176
		176

To show the area of a circle, and its square: Cut a piece of wire 38 inches long, form a circle, and the diameter will be 12 inches, and its area will be 112. Cut another piece 38 inches long, form a perfect square, whose sides will be $9\frac{1}{2}$ inches each, and whose area will be $90\frac{1}{4}$ inches. You will see by this the difference of square and circular measure. The same length of wire formed into a circle measures in area 112 inches, and formed into a square, $90\frac{1}{4}$; making a difference of $21\frac{3}{4}$ inches in the shape of the circle and square.

There is a fish whose tail weighs 9 pounds; his head weighs as much as his tail and half his body; his body weighs as much as his head and tail; what is the weight of the fish?

Answer: His head weighs 27 pounds, his body 36 pounds; the whole fish, 72 pounds.

To measure a coal pit of wood: Suppose your pit is 60 feet in circumference; first find the diameter; and to find it divide the circumference by 19, and multiply the product by 6; this gives the diameter. Then find the area, by taking $\frac{7}{9}$ of the diameter; that is the square of the diameter. Then multiply by the mean height, which will be 6 feet, as the pit is 12 feet high; then you have the cubic feet. Then divide by the number of feet in a cord, which is 128 feet, which gives the cords, feet, and inches.

19) 60 feet, circumference.

3.157
6

18.942

So call the diameter 19 inches.

19
19

171
19

361 This is the square.
7

I take $\frac{7}{9}$ of the square. 9) 2527

280

This is $\frac{7}{9}$. I multiply the 280 by 6, half the height.
6

1680

I divide this by 128, the feet in a cord.

```
128 ) 1680 ( 13 cords, 16 feet.
      128
      ---
       400
       384
       ---
        16
```

Suppose your pit is 60 feet in diameter.

```
         60
         60
       ----
       3600
          7
      -----
  9 ) 25200
      -----
       2800
          6
      -----
128 ) 16800 ( 131 cords, and 32 feet.
```

THE SQUARE OF THE CIRCLE.

The square of a circle 12 inches in diameter is $9\frac{1}{2}$ inches. This is the square of the circle, but is not the square of the area of the same circle.

The square of the area of a circle 12 inches in diameter is 10 inches and a half, and $\frac{8}{10}$, and a little over, thus:—

```
         112 ( 10.583
         1
        ----
  205 ) 01200
         1025
        -----
 2108 ) 17500
         16864
        ------
21163 ) 63600
         63489
        ------
           111, remainder.
```

Thus you see it is 10 inches, $\frac{1}{2}$, $\frac{8}{10}$, and $\frac{3}{100}$.

The fourth part of a circle is the square of the circle; that is, the square of the circle itself, but not the area. Thus:—

```
4 ) 38 ( 9.5
    36
    ——
     20
     20
     ——
      0
```

The circle gains over all other measure. Take a circle 38 inches round, which is 12 diameter, and the area of this circle is 112 inches. Now take a square 38 inches round, that is, whose four sides measure 38 inches, and the area will be 90$\frac{1}{4}$ inches. 12 diameter is 112 area, 6 diameter is 28 area. It varies the same as the square; so that $\frac{1}{2}$ the diameter is $\frac{1}{4}$ in area.

Suppose a circle 12 inches in diameter to be filled with rings $\frac{1}{6}$ of an inch wide. Now, the length of the average ring is 19 inches. Take 36 rings, 19 inches long and $\frac{1}{6}$ of an inch wide, and measure them by square measure, and you will make them measure 114 inches; but by circular measure you find they measure 112 inches. Now take 36 rings, 19 inches long and $\frac{1}{6}$ of an inch wide, and by square measure (that is, 90 degrees) each ring will measure 3 inches and $\frac{1}{6}$, which will make the 36 rings measure 114 inches; but by circular measure, (that is, in proportion of 60 degrees,) each ring will measure 3 inches and $\frac{1}{9}$, and the 36 rings will measure 112 inches.

This shows the difference between circular measure and square measure. It is in one ring $\frac{1}{18}$ of an inch, which is circular measure, in proportion of 90 degrees, which is to the square as 56 to 57; for twice 57 are 114, and twice 56 are 112. The present measure is as 132, and correct measure is as 133. As 7 to 22 would be 132, and as 6 to 19 would be 133. The difference between $\frac{1}{6}$ and $\frac{1}{9}$ is $\frac{1}{3}$.

In regard to the taking of levels, you will find that great

errors are made, for the want of a perfect quadrature of the circle. Thus I have frequently heard it said, if you make a platform perfectly straight and level, and place a perfectly round ball on it, it would make a perpetual motion. But they do not consider, I think, that the creation is on a perfect circle, therefore a perfect level and a perfect straight cannot exist together in any one thing. In 16 miles it will get 56 yards out of level; in 8 miles, 14 yards; in 4 miles, 3½ yards; in 2 miles, 2 feet 7 inches; in one mile, 7¾ inches; in half a mile, 1⅞ inches; in ¼ of a mile, ⅜ of an inch.

On a survey of the Isthmus of Darien, to consider the practicability of cutting a passage through from ocean to ocean, to facilitate the navigation of the world, the return was, that one ocean was higher than the other, and therefore it would be impossible to cut such a passage. I think this return was wrong, for the want of the perfect quadrature of the circle; for reason must dictate that these great bodies of water would naturally find their level, except agitated by winds, or some natural causes.

In taking a level for a railroad, they may take a spirit level, and take a sight at a great distance, which will be imperfect, if they do not calculate for the curve of the circle.

I was asked by a man to give a plan to find a proportion for stove pipes, and other things of the kind, such as cabooses, &c.; for which I take the rule of three, or proportion, thus: Suppose pipe to be square, say 6 inches; then $6 \times 6 = 36$. Now, the circle is to the square as 7 to 9; so, by the rule of three,—

$$\begin{array}{r} \text{If } 9 : 36 : : 7 \\ 7 \\ \hline 9\,)\,252 \\ \hline 28\text{, answer.} \end{array}$$

Any other measure that may be required in your business will work in the same way. $12 \times 12 = 144$; so,—

If 9 : 144 : : 7
7

9) 1008

112, answer.

Thus you see the proportion may be got in a circle as perfectly as in a square. Or suppose it is an oval; then find the measure of the circle, as above, say 6 inches diameter: —

If 9 : 36 : : 7
7

9) 252

28

But say the oval is $\frac{1}{4}$ longer; so 6 diameter, $7\frac{1}{2}$ long; the area of the circle is 28; so 4 times 7 are 28, which added to 7, = $\frac{1}{5}$, gives 35 inches for the area of the oval.

The question is often asked, why a polygon of four sides will not measure a circle. There is no figure, of any number of sides, that will measure the circle, but the hexagon, or figure of six sides. This is the only figure that is equal in all its parts. The hexagon is from an angle of 60 degrees; square measure is from an angle of 90 degrees; these angles differ 30 degrees.

With the same propriety it might be asked, why will not an angle of 60 degrees measure a square. I not only say there is but one figure that will measure a circle, but I also have but one number that will measure a circle; and that is the number 6. With these — the hexagon and the number 6 — I have perfected the measure of the circle.

If you take a trigon, of 3 sides, or a tetragon, of 4 sides, or a pentagon, of 5 sides, or a heptagon, of 7 sides, or an octagon, of 8 sides, or a nonagon, of 9 sides, or a decagon, of 10 sides, or an endecagon, of 11 sides, or a dodecagon, of 12 sides, either of these polygons would, in the operation, produce the surd number, and that would produce fractions, which I think no man would comprehend, as it would lead to the same difficulty which has been felt in all ages.

Time is prefigurative of the number 6; and as time is measured by a circle, I take the number 6 to measure the circle; and the only polygon that is synonymous with that number is the hexagon. I am of the opinion that there is no other number but the number 6 whereby the measure of the circle could have been effected. I should not have taken the number 6 from choice, but from necessity, for 10 would have been preferable; but 6 was the only number whereby I could measure it without surd numbers; and I should think no man would prefer irrational numbers to work out a difficult problem, when he could have rational ones. The hexagon, or polygon of 6 sides, is the only one whereby I could get every angle, every side, and every measure perfect; and if the foundation is imperfect, so would be the superstructure.

Now, $6 \times 6 = 36$, equal the 2d power; and $36 \times 36 = 1296$, or the 4th power, which is the square of the square; and I can make nothing of the square of the square but the circle.

The same question might be asked, why will not a tetrahedron, of 4 sides, or an octahedron, of 8 sides, or a dodecahedron, of 12 sides, or an icosahedron, of 20 sides? If any of these angles would have measured the circle, it is more than probable it would have been done years ago.

THE NUMBER SIX.

As I have before said, for my prime number in the measure of the circle I have taken the number 6. For this I refer to Plato, who says time is prefigurative of the number 6. From this observation of Plato's, our ancestors have thought the world was to last but 6000 years. In the 3d chapter of the 2d epistle of Peter the same opinion is confirmed. You there read of the scoffers and unbelievers; and St. Peter's answer was,

"Be not ignorant of this one thing; that one day is with the Lord as a thousand years; for slackness is not counted to the Lord as it is to man." St. Peter says it certainly will come. Now, as the world was 6 days in making, I draw the inference that it will stand 6000 years. This was the opinion of B. P. Burket, and many of our ancestors who were considered wise and great men.

You may observe my ratio to measure a circle (which is a true figure of eternity) would naturally end with a 6; and thus you see a true ratio never could have been found by the natural use of mathematics. In the Revelation of St. John you find that after the 7th seal the world is measured to its end by 6 trumpets. Now, under 6 trumpets you find the four angels loosed from the great river Euphrates, by whose army the greater part of men were slain. This army was computed at 200,000,000. The oath in chapter 10 was, that when the 7th trumpet began to sound, the end of the world should come; so you see it lasts but through the 6. And in the 13th chapter you find, under the reign of the great beast, that no one had liberty to buy or sell except he had the mark of the beast, or the number of his name. Let him that hath understanding count his number; and his number is 666. Now, it goes by thirds, as you may see by reading; as 3 times 3 are 9, and 666 multiplied by 9 is to the square equal to 5994. This makes it lack 6 of making the 6000 years. In the 24th chapter of Matthew our Savior says, "Except these days be shortened, no flesh shall be saved." Now, the number of the beast falls short of 6000 by 6 years; so it appears that the time was shortened 6 years; and you see that 6 is the number that governs all these operations; and it is evident that great operations can be performed with it.

Now, I offer as proof of the number 6: The number 6 is not proficient of the measure of the circle, therefore I go to the number 7; and what I take of the number 7 is synonymous with the lack of the 6000 years, which is my biquadrate, and is what makes up the measure of the circle.

Now, the hexagon has 6 sides, 6 squares, 6 angles, and 6 radii; the 6 radii are each 6 inches in length, the 6 squares each 6 inches in length, and the 6 angles each 6 inches in length. The 6 sides, each 6 inches, the 6 oblong squares, with the biquadrate added, is the perfect measure; as 6 times 6 are 36, and $\frac{1}{3}$ of 6 added to 6 times 6 is 38, which is the complete and scientific mathematical measure of the circle.

Time was to be continued through the sounding of the 6 trumpets; and when the 7th began to sound, time was to be no more. Now the want of the measure of the circle is no more when I take of the number 7 my biquadrate. As I believe the measure of the circle was not to be effected until the sounding, or fulfilment of the Scriptures, and as it is governed by the number 6, it was not to be measured in the sounding of the 1st, 2d, 3d, 4th, or 5th trumpet, but in the sounding of the 6th.

TRIGONOMETRY.

I am told that trigonometry has found the measure of the circle to a hair's breadth. What is meant by it I do not know, nor can I conceive. Trigonometry cannot compute the true perimeter of a circle, unless that circle be already known.

Most men are aware that computation is governed by a circle; they also believe — for they find it in practice — that there is a truer measure than the one now in use; and if all measurement is governed by a circle, it comprehends a gallon as well as a bushel, and a right line as well as a curve; and if the circle which directs these measures be incorrect, it is easy to see why all offsprings of that circle should be deficient in the same proportion. If we wish to compute a segment of a circle, we take spherical trigonometry to do so; but we must understand at the same time that trigonometry cannot give us the truth, since we know its foundation is affected more or less by an error known by mathematicians as most dangerous for its counterbalance, wherein it contracts truth, shows truth, proves truth, and is still a great falsehood. We must therefore be acquainted with the true circle before we can find its true parts; and as trigonometry is but that portion of science which executes the laws of geometry, and as geometry has never declared a circle to be equal to any figure but itself, we cannot take trigonometry to give us a segment, or a whole circle, within a hair's breadth, or any other breadth, with confidence.

I am told that the pendulum swings a perfect yard. This I do not understand. It must take a very ingenious mechanic to make a pendulum swing a perfect yard, when no man as yet but myself ever knew what length a perfect yard was.

If we admit that the length of a yard was known to man, it would be impossible, from the nature of things, to make a pendulum that would swing a perfect yard, because the influence of the atmosphere upon the metal would cause it to vary more from correct measure than as 7 to 22 does. It has not as yet been ascertained that any composition of metals can be so worked as to obviate the variation caused by atmospheric influence; so that this doctrine is much like that of the gentleman professor in England who told me that if I worked a problem right to measure a circle, there would always be a remainder; therefore, in order to do a thing right, I must do it wrong; for my work was not mathematical because it came right. This is a strange doctrine, although it is believed by a great many in this world.

QUESTIONS ASKED.

I am told by the learned classes that the circle is measured as near as it ever can be, and near enough for any thing. I will ask a number of questions, which I think will illustrate the work to the scholar: —

1. From what do all measures spring, or from what derived?
2. From what do all weights spring, or from what derived?
3. What makes a mathematical inch? how do you arrive at this?
4. What makes a pound or an ounce? how can you tell?

5. What is the measure of the circle, and what is the use of it?

6. Why has it remained in oblivion if it is useful to man?

7. How do you find the circumference of a circle?

8. How do you find the area of a circle?

9. Having the circumference, how do you find the diameter?

10. Having a promiscuous number of figures for an area, how do you find the circle that bounds them?

11. What is the 4th power, or biquadrate?

12. What part of a circle is the biquadrate?

13. What are the square inches of the dodecagon?

14. What makes up the measure of the circle?

15. What is the square of the circle?

16. What will be the size of a circle from the square of 9½ inches?

17. What will be the size of the square from a circle 12 inches in diameter?

18. What is the difference between square and circular measure?

19. What angle is square measure from?

20. What angle is circular measure from?

21. What is the difference between the two angles?

22. What measure is that which gains over all others?

23. What proportion is the square to the circle?

24. How much does the circle gain over the square?

25. How do you measure a cone?

26. How do you measure a cylinder?

27. How do you find a level on the earth?

28. How do you find the curve of the earth?

29. What is the curve of the globe?

30. What proportion has the diameter of a circle to its circumference?

31. What is the difference between as 7 to 22 and as 6 to 19?

32. How much is long measure astray?

33. How much is square measure astray?

34. How much is cubic measure astray?
35. How much do 1728 cubic inches of water weigh?
36. How many cubic inches are there in a bushel?
37. What portion of 1728 cubic inches is one ounce?
38. What portion of 1728 cubic inches is one pound?
39. What is time measured by?
40. Why does the number 6 measure time?
41. Why does the number 6 measure a circle?
42. Why is the hexagon the only figure or angle to measure a circle?
43. What is a surd number?
44. How do you extract the square root of the surd number?
45. What makes a surd number?
46. What portion of a cylinder, supposing it to be square, to find the measure in square inches, is the circle of it?
47. How do you find the square inches in a cylinder, and by what measure?
48. How do you measure a grindstone?
49. How do you measure land, if in a circular form?
50. By what rule do you find the square inches in a steam boiler?
51. By what rule do you find the square inches in a globe?
52. By what rule do you measure a cask correctly?
53. Can square measure perfect circular?
54. Can circular measure perfect square measure?
55. What measure has been sought for in all ages of the world?
56. What measure was perfected in 1845?
57. What does the angle gain on the circle?
58. What is three times the square of the radius in a circle 12 inches in diameter?
59. Why does the circle measure the area?
60. Why does the area measure the circle?
61. Why does the square prove the circle?
62. Why does the circle prove the square?
63. Why does the square prove the area?

64. Why does the area prove the square?

65. Suppose you form a circle from a piece of wire 38 inches long, and a square from a piece 38 inches long; what is the difference of the area and the square?

66. Why is one half the diameter one fourth in area?

67. What is the square root of a circle 12 inches in diameter?

68. What is the difference between the square of the circle and the square of the area?

69. What square will a circle form that is 12 inches in diameter?

70. What circle will a square make that is $9\frac{1}{2}$ inches square?

71. What circle is that which has 112 for area?

72. What area is that which has a circle 12 inches in diameter?

73. What area has a circle 17 inches in diameter?

74. What area has a circle that has a radius of 16 inches?

75. What would be the radius of a circle that has 224 inches area?

76. What is the difference of area between a circle 12 inches in diameter, and one 6 inches in diameter?

77. What is the difference in the length of a string that goes round the circle, and one that goes round the square?

78. How do you find the difference of area between a circle and a square?

79. What is the difference of area of any circle derived from as 7 to 22, or as 6 to 19?

80. Suppose you travel on a radius 100 miles; what is the difference of circular measure from as 6 to 19 or as 7 to 22?

81. Suppose you shot a ball level on the earth 1000 miles, and it then fell perpendicularly to the earth; how far would it have to fall?

82. Suppose your radius is 100 miles; how much larger is your circle?

83. Suppose your radius is 40 miles; what is the area of the circle?

84. Suppose your circle is 4 times the radius; what is the area?

85. Suppose one third of your radius is 12 inches; what is the circumference of your circle?

86. What is the biquadrate when your radius is 12 inches?

87. What is the biquadrate when the radius is $3\frac{1}{2}$ inches?

88. What is the biquadrate when your circle is 78 inches?

89. What is the biquadrate when the area is 240 inches?

90. What is the biquadrate of a circle which has an area of 17 inches?

MEASURES FOR GRAIN, ETC.

1. A measure to hold a bushel, if in circular form, must be 18 inches in diameter, and 9 inches, lacking $\frac{21}{100}$ of an inch, in depth.

2. A measure to hold half a bushel, if in circular form, must be $12\frac{3}{4}$ inches in diameter, and 9 inches deep.

3. A measure to hold a peck, if in circular form, must be 10 inches in diameter, and $7\frac{1}{4}$ inches deep.

4. A measure to hold 4 quarts, if in circular form, must be $8\frac{1}{2}$ inches, lacking $.019\frac{4}{9}$, in diameter, and must be 5 inches deep, which is a little too much, as 280 to 282.

5. A measure to hold 2 quarts, if in circular form, must be 6 inches in diameter, and $5\frac{1}{28}$ inches deep.

6. A measure to hold 1 quart, if in circular form, must be $4\frac{1}{4}$ inches in diameter, and 5 inches in depth.

7. A measure to hold a pint, if in circular form, must be 3 inches in diameter, and 5 inches deep.

8. A measure to hold half a pint, if in circular form, must be $2\frac{1}{2}$ inches in diameter, and $3\frac{6}{10}$ inches deep.

9. A measure to hold a gill, if in circular form, must be 2 inches in diameter, and 3 inches deep.

10. A measure to hold a glass, if in circular form, must be $1\frac{1}{2}$ in diameter, and $2\frac{1}{2}$ inches deep.

On the following page is a table showing what length to cut iron that is used for hoops, wagon tires, and all such uses as require a length to be cut for a circumference. Nothing is allowed for drawing; I allow the iron to be half an inch thick. I calculate from a diameter of 6 feet to one of 1 foot:—

Length.		Circumference.		Length of hoop.		Length.		Circumference.		Length of hoop.	
Feet.	In.	Feet.	Inches.	Feet.	Inches.	Feet.	In.	Feet.	Inches.	Feet.	Inches.
6	0	19		18	$11\frac{1}{6}$	3	5	10	$9\frac{5}{6}$	10	$10\frac{3}{8}$
5	11	18′	$8.8\frac{5}{6}$	18	$8\frac{5}{6}$	3	4	10 ·	$6\frac{4}{6}$	10	$7\frac{5}{6}\ \frac{1}{18}$
5	10	18	$5.4\frac{4}{6}$	18	$4\frac{5}{6}\ \frac{1}{18}$	3	3	10	$3\frac{3}{6}$	10	$4\frac{2}{6}\ \frac{2}{18}$
5	9	18	$2\frac{3}{6}$	18	$1\frac{4}{6}\ \frac{2}{18}$	3	2	10	$0\frac{2}{6}$	10	$1\frac{1}{6}$
5	8	17	$11\frac{2}{6}$	17	$10\frac{4}{6}$	3	1	9	$9\frac{1}{6}$	9	$10\frac{1}{6}\ \frac{1}{18}$
5	7	17	$8\frac{1}{6}$	17	$7\frac{3}{6}\ \frac{1}{18}$	3	0	9	6	9	$7\ \frac{2}{18}$
5	6	17	5	17	$4\frac{2}{6}\ \frac{2}{18}$	2	11	9	$2\frac{5}{6}$	9	4
5	5	17	$1\frac{5}{6}$	17	$1\frac{2}{6}$	2	10	8	$11\frac{4}{6}$	9	$0\frac{5}{6}\ \frac{1}{18}$
5	4	16	$10\frac{4}{6}$	16	$10\frac{4}{6}\ \frac{1}{18}$	2	9	8	$8\frac{3}{6}$	8	$9\frac{4}{6}\ \frac{2}{18}$
5	3	16	$7\frac{3}{6}$	16	$7\ \frac{2}{18}$	2	8	8	$5\frac{2}{6}$	8	$6\frac{4}{6}$
5	2	16	$4\frac{2}{6}$	16	4	2	7	8	$2\frac{1}{6}$	8	$3\frac{2}{6}\ \frac{1}{18}$
5	1	16	$1\frac{1}{6}$	16	$\frac{5}{6}\ \frac{1}{18}$	2	6	7	11	8	$0\frac{2}{6}\ \frac{2}{18}$
5	0	15	10	15	$9\frac{4}{6}\ \frac{2}{18}$	2	5	7	$8\frac{5}{6}$	7	$9\frac{2}{6}$
4	11	15	$6\frac{5}{6}$	15	$6\frac{3}{6}$	2	4	7	$5\frac{4}{6}$	7	$7\frac{1}{6}\ \frac{1}{18}$
4	10	15	$3\frac{4}{6}$	15	$3\frac{3}{6}\ \frac{1}{18}$	2	3	7	$2\frac{3}{6}$	7	$4\ \frac{2}{18}$
4	9	15	$0\frac{3}{6}$	15	$0\frac{2}{6}\ \frac{2}{18}$	2	2	6	$11\frac{2}{6}$	7	1
4	8	14	$9\frac{2}{6}$	14	$9\frac{1}{3}$	2	1	6	$8\frac{1}{6}$	6	$9\frac{5}{6}\ \frac{1}{18}$
4	7	14	$6\frac{1}{6}$	14	$6\frac{1}{6}$	2	0	6	4	6	$6\frac{4}{6}\ \frac{2}{18}$
4	6	14	3	14	$3\ \frac{2}{18}$	1	11	6	$0\frac{5}{6}$	6	$2\frac{4}{6}$
4	5	13	$11\frac{5}{6}$	14	0	1	10	5	$9\frac{4}{6}$	5	$11\frac{3}{6}\ \frac{1}{18}$
4	4	13	$8\frac{4}{6}$	13	$8\frac{5}{6}\ \frac{1}{18}$	1	9	5	$6\frac{3}{6}$	5	$8\frac{2}{6}\ \frac{2}{18}$
4	3	13	$5\frac{3}{6}$	13	$5\frac{4}{6}\ \frac{2}{18}$	1	8	5	$3\frac{2}{6}$	5	$5\frac{1}{6}$
4	2	13	$2\frac{2}{6}$	13	$2\frac{4}{6}$	1	7	5	$0\frac{1}{6}$	5	$2\frac{1}{6}\ \frac{1}{18}$
4	1	12	$11\frac{1}{6}$	12	$11\frac{3}{6}\ \frac{1}{18}$	1	6	4	9	4	$11\ \frac{2}{18}$
4	0	12	8	12	$8\frac{2}{6}$	1	5	4	$5\frac{5}{6}$	4	8
3	11	12	$4\frac{5}{6}$	12	$5\frac{2}{6}$	1	4	4	$2\frac{4}{6}$	4	$4\frac{5}{6}$
3	10	12	$1\frac{4}{6}$	12	$2\frac{1}{6}\ \frac{1}{18}$	1	3	3	$11\frac{3}{6}$	4	$1\frac{5}{6}\ \frac{2}{18}$
3	9	11	$10\frac{2}{6}$	11	$11\ \frac{2}{18}$	1	2	3	$8\frac{2}{6}$	3	$10\frac{4}{6}$
3	8	11	$7\frac{2}{6}$	11	$7\frac{5}{6}$	1	1	3	$5\frac{1}{6}$	3	$7\frac{3}{6}\ \frac{1}{18}$
3	7	11	$4\frac{1}{6}$	11	$4\frac{5}{6}\ \frac{1}{18}$	1	0	3	2	3	$4\frac{2}{6}\ \frac{2}{18}$
3	6	11	$1\frac{5}{6}$	10	$11\frac{4}{6}\ \frac{2}{18}$						

To find the height of a steeple: Take a quadrant, and place yourself a distance of 45 degrees from the base; that is, place yourself at such a distance as to make the quadrant strike the top of the steeple; then measure from the base to where you stand, and allow for your height, and that will be the height of the steeple.

To show the loss on linear, square, and cubic measure:

```
            2 ) 132 ( 66                    3 ) 132 ( 44
If 132 : 1 : : 100          If 66 : 1 : : 100   If 44 : 1 : : 100
         100                          100                 100
         ———                          ———                 ———
 132 ) 1000 ( .76 per cent.  66 ) 100 ( 1.51    44 ) 100 ( 2.28
        924                         66                  88
        ———                         ———                 ———
         76                         340                 120
                                    330                  88
                                    ———                 ———
                                    100                 320
                                     66                 352
                                    ———                 ———
                                     34              little short.
```

So it makes linear measure near or about 76 per cent.; square measure, about 1.51 per cent.; and cubic measure about 2.28 per cent. But as you cannot use the angle in the work, it gives too little in cubic measure, and weights are in proportion of that measure. Now, the statute of America says that a gallon of water weighs just 8 pounds, and 221 inches make a gallon, and 1 cubic foot of water weighs 62½ pounds; so, as measurements of all kinds are wrong, so are weights; and weights are wrong in proportion of cubic measure.

In England, the wine gallon is 231 inches, which is 10 inches more than it is in America; and yet they reckon 8 pounds to the gallon, which shows their weights and measures to be about $\frac{1}{23}$ part more than the American weights and measures, which accounts for the remark common in England when I was a boy, that when any thing was short of measure it was Presbyterian measure; the reason for which was, I presume, that the Presbyterians came to America to enjoy their religious freedom, and made their measures a little less; so the English called it Presbyterian measure. The foot and inch, however, are the same in both countries. The English call 231 inches a gallon, and the Americans 221; but the right measure would be 226; so the English gallon is 5 inches too much, and the American 5 too little; all weights, therefore, should be altered in the same proportion.

NUMERATION.

It is seldom that persons numerate more than 12 figures. I saw a plan in New Haven, Connecticut, which was new to me; it ran to 10 figures, for millions, and after that every denomination went by 3 figures, thus: Billions, tens of billions, hundreds of billions; trillions, tens of trillions, hundreds of trillions, &c.

But I think the way is to numerate by logarithms, thus: 3 figures for hundreds, 6 for thousands, 12 for millions, 24 for billions, 48 for trillions, 96 for quadrillions, &c.; there would then be no number that could not be numerated; if there were 100 figures I could numerate them, but I never found another that could.

PERMUTATION.

The English language has 20,500 nouns, 40 pronouns, 9,500 adjectives, 8,000 verbs, 2,600 adverbs, 69 prepositions, 19 conjunctions, 68 interjections, and 2 articles; making in all 40,000 words in the language.

The alphabet has 26 letters. I have worked to find how many different changes there are in the English alphabet, and find them to be 4,226,579,232,623,185,627,342,000,000. I numerate this thus: 4226 trillions, 579,232 millions of billions, 623,185 billions, 627,342 millions.

These are the different ways that the 26 letters are changed to make words and representations. In this numeration there are two ciphers too many.

NUMERATION TABLE.

0	Units.
0	Tens.
0	Hundreds.
0	Thousands.
0	Tens of thousands.
0	Hundreds of thousands.
2	Millions.
1	Tens of millions.
5	Hundreds of millions.
9	Thousands of millions.
1	Tens of thousands of millions.
7	Hundreds of thousands of millions.
6	Billions.
5	Tens of billions.
4	Hundreds of billions.
3	Thousands of billions.
1	Tens of thousands of billions.
2	Hundreds of thousands of billions.
1	Millions of billions.
9	Tens of millions of billions.
8	Hundreds of millions of billions.
7	Thousands of millions of billions.
6	Tens of thousands of millions of billions.
5	Hundreds of thousands of millions of billions.
4	Trillions.
3	Tens of trillions.
2	Hundreds of trillions.
1	Thousands of trillions.
4	Tens of thousands of trillions.
3	Hundreds of thousands of trillions.
2	Millions of trillions.
1	Tens of millions of trillions.
1	Hundreds of millions of trillions.
9	Thousands of millions of trillions.
8	Tens of thousands of millions of trillions.
7	Hundreds of thousands of millions of trillions.
6	Billions of trillions.
5	Tens of billions of trillions.
4	Hundreds of billions of trillions.
3	Thousands of billions of trillions.
3	Tens of thousands of billions of trillions.
2	Hundreds of thousands of billions of trillions.
2	Millions of billions of trillions.
1	Tens of millions of billions of trillions.
8	Hundreds of millions of billions of trillions.
7	Thousands of millions of billions of trillions.
6	Tens of thousands of millions of billions of trillions.
5	Hundreds of thousands of millions of billions of trillions

How many changes can be rung on 12 bells?

1	1
	2
2	2
	3
3	6
	4
4	24
	5
5	120
	6
6	720
	7
7	5040
	8
8	40320
	9
9	362880
	10
10	3628800
	11
	3628800
	3628800
11	39916800
	12
	79833600
	39916800
12	479001600

MULTIPLICATION

Is an expeditious way of performing several additions, as in this case: Required to multiply £1 19 s. 11 d. 3 qr. by itself; what will be the answer? Ans. £3 19 s. 11 d. 2 qr.

THE MILE IN DIFFERENT COUNTRIES.

The English mile contains	1760 yards.
The Russian mile contains	1100 yards.
The Irish and Scotch mile contains	2200 yards.
The Staton mile contains	1460 yards.
The Polish mile contains	4400 yards.
The Spanish mile contains	5028 yards.
The German mile contains	5866 yards.
The French league contains	3666 yards.

Algebra is the art and science of some ingenious, cunning thoughts, calculated to fill the scholar's mind, and take his labors for foolish employment, which will do him no good, but darken his literary prospects. I would advise all to have nothing to do with it, as I consider it useless to all science and literature. It is a labored, worthless study, calculated by craft, for the sake of fame. It gives the scholar much labor and anxiety, which keeps his mind from other useful and important sciences.

Doctrine of annuities is to find the value of any annuity.

Logarithms are certain artificial numbers.

Trigonometry is finding all the sides and angles of a triangle.

Astronomy is the science of the heavenly bodies.

Mechanical powers are six in number.

Electricity is combustible air, such as lightning.

Medical electricity is applying it to different uses.

Pneumatics is that part of natural philosophy that treats of the weight, pressure, and electricity of air.

Hydrostatics treats of the equilibrium of fluids, or the gravitation of fluid bodies.

All fluids except air are incompressible.

Fluidity, according to Sir Isaac Newton's definition, is the yielding of a body when compressed by any force, the parts of the body being easily moved.

A quart of air weighs 16 grains.

Hydraulics is the science of the force and motion of fluids.

Geometrical progression is when the numbers increase thus: 1, 2, 4, 8, 16, 32, 64, 128, &c.

Arithmetical progression is when the numbers increase thus: 3, 6, 9, 12, 15, 18, 21, 24, 27, 30, 33, 36, 39.

The number of inches in an English gallon is, for oil, 282, for wine, 231. In America it is 221 inches. In a bushel there are 2256 inches; in half a bushel, 1128; in one peck, 564; in one gallon, or half a peck, 282; in one quart, $70\frac{1}{2}$; in one pint, $35\frac{1}{4}$; in half a pint, $17\frac{6}{10}$ and $\frac{1}{4}$; in a gill, $8\frac{1}{2}$; in a glass, $4\frac{1}{2}$; in half a glass, $2\frac{1}{8}$.

Wightman, in his arithmetic, says there are 2218 inches in a bushel, and he also says it measures in diameter $19\frac{1}{2}$ inches, and is $18\frac{1}{4}$ inches deep. Now, according to this he would have 2441 inches in a bushel.

Thomas Hodson says, in his "Tutor," that there are 282 inches in an oil gallon, and he makes a bushel to contain 2256 inches, that is, if you take the oil measure; and if we take our statute gallon of 221 inches, the bushel will be 1768 inches.

Now, which of these is right is not for me to say; I leave that for those who may be interested.

Euclid, the celebrated mathematician, lived 277 years before Christ.

Archimedes lived 208 years before Christ.

Ptolemy lived 148 years before Christ.

John Napier, of Scotland, lived in 1600, and discovered logarithms.

James Gregory lived in 1675. He was also of Scotland.

Isaac Barron lived in 1677. He was also of Scotland.

John Flamsteed lived in 1719. He was of Derbyshire, in England.

Dr. John Kell lived in 1719.

Sir Isaac Newton lived in 1700. He was of Lincolnshire.

The above were celebrated mathematicians.

Pythagoras, one of the greatest philosophers, considered mathematics of great importance to man. He says mathematics is the first step towards wisdom. Lord Bacon says men do not sufficiently understand the value and importance of mathematics. It is a noble and divine system, calculated for the benefit and enterprise of mankind generally.

GEOMETRY.

There are three different angles — the right angle, the obtuse angle, and the acute angle.

When a plain figure is terminated by three lines it is called a triangle.

There are three particular sorts of triangles — equilateral, isosceles, and right angled.

When the three sides of a triangle are equal, it is called an equilateral triangle.

When two sides only are equal it is called an isosceles triangle.

When one angle is a right angle, it is called a right angled triangle.

Any side of a triangle may be called the base, and the angle opposite to it is called the vertex.

The side opposite to a right angle is called the hypotenuse

LINEAR MEASURE.

12 inches make 1 foot.
3 feet " 1 yard.
6 feet " 1 fathom.
16½ feet, or 5½ yards, make 1 pole or rod.
40 poles make 1 furlong.
8 furlongs " 1 mile.

SQUARE MEASURE.

144 inches make 1 foot.
9 feet " 1 yard.
36 feet " 1 fathom.
272¼ feet, or 30¼ yards, make 1 rod.
1600 poles make 1 furlong.
64 furlongs " 1 mile.

160 rods make 1 acre.
80 rods equal ½ an acre.
40 rods " ¼ of an acre.
20 rods " ⅛ of an acre.
10 rods " $\frac{1}{16}$ of an acre.
5 rods " $\frac{1}{32}$ of an acre.
640 acres make 1 square mile.
144 square inches make 1 square foot.
1728 cubic inches make 1 cubic foot.

On the opposite page will be found the compound interest of one dollar, from 1 year to 12, at 6 per cent.

One dollar, at 6 per cent., will double in 12 years; and the simple interest of one dollar for 12 years would be 72 cents; so the compound interest of one dollar for 12 years is 28 cents.

1

6

6, 1 year.

100

1.06

6

6.36

106

1.1236, 2 years.

6

6.7416

112.36

1.191016, 3 years.

6

7.146096

1.191016

1.26247696, 4 years.

6

7.57486176

126.247696

1.3382255776, 5 years.

6

8.0293534656

1.3382255776

1.418519112256, 6 years.

6

8.511114673536

1.418519112256

1.50363025899136, 7 years

6

9.02178155394816

1.50363025899136

1.5938480745308416, 8 years.

6

9.5630884471850496

1.5938480745308416

1.688478958902692096, 9 years.

6

10.130873753416152576

1.688478958902692096

1.78978769643685362176, 10 years.

6

10.7387261786211217 3076

1.7897876964368536 2176

1.88717495822306483 90676, 11 years.

6

11.3230497493383890344056

1.8871749582230648390676

2.0004054557164487294116 56, 12 years.

The interest of one dollar for 12 years, at simple interest, would be 72 cents.

1.00

6

6.00

100

106.00

12

$1.7200

Compound, 28

$2.00

The manifestation of the spirit is given to every man, to profit withal. To one is given wisdom; to another is given knowledge; to another is given, faith; to another, the gift of healing; to another, the working of miracles; to another, prophecy; to another, discerning of spirits; to another, divers kinds of tongues; to another, interpretation of tongues.

But all these work that selfsame spirit, dividing to every man severally as he will.

A TABLE OF GAUGING.

I find the mean diameter by adding the bung and head together; one half of this will be the mean diameter. Multiply by the length, which gives the contents, in cubic inches; divide by 221 inches, the number of inches in a gallon, as this table is calculated, which gives the gallons. The table is laid down in columns every inch in length, with decimals if the length is short or long, for the diameter laid down; one statement in the rule of three will find it, or you may find how much it is too long or short by tabular measure. Suppose it is 30 inches long in the table, and cask 32; then your cask will hold 15 gallons more.

Mean diameter.	Length.	Gallons	Mean diameter.	Length.	Gallons.	Mean diameter.	Length.	Gallons.
39.9	48	269.7	38.4	46	238.8	36.9	44	210.95
39.8	48	267.7	38.3	46	237.5	36.8	44	209.8
39.7	48	266.38	38.2	46	236	36.7	44	208.67
39.6	48	265.04	38.1	46	235	36.6	44	207.35
39.5	48	263.7	38	46	233.8	36.5	44	206.4
39.4	47	256.9	37.9	46	232.65	36.4	44	205.27
39.3	47	255.6	37.8	45	226.4	36.3	44	204.14
39.2	47	254.25	37.7	45	225.2	36.2	44	203
39.1	47	253	37.6	45	224	36.1	44	201.9
39	47	251.7	37.5	45	222.8	36	44	200.7
38.9	47	250.42	37.4	45	221.6	35.9	43	195
38.8	46	243.8	37.3	45	220.5	35.8	43	194
38.7	46	242.5	37.2	45	219.7	35.7	43	192.9
38.6	46	241.3	37.1	45	218.09	35.6	43	191.9
38.5	46	240.08	37	45	216.9	35.5	43	190.8

TABLE OF GAUGING, CONTINUED.

Mean diameter.	Length.	Gallons.	Mean diameter.	Length.	Gallons.	Mean diameter.	Length.	Gallons.
35.4	43	189.7	31.1	37	126	26.9	34	86.5
35.3	43	188.66	31	37	125.2	26.8	34	85.9
35.2	43	187.6	30.9	36	121	26.7	34	85.3
35.1	43	186.5	30.8	36	120.25	26.6	34	84.7
35	43	185.4	30.7	36	119.7	26.5	34	84
34.9	42	180.2	30.6	36	118.66	26.4	34	83.29
34.8	42	179.3	30.5	36	117.9	26.3	34	82.8
34.7	42	178.4	30.4	36	117.1	26.2	34	82.14
34.6	42	177.5	30.3	36	116.3	26.1	34	81.55
34.5	42	176.6	30.2	36	115.6	26	34	80.9
34.4	41	170.7	30.1	36	114.49	25.9	33	77.94
34.3	41	169.85	30	36	114	25.8	33	77.33
34.2	41	168.29	29.9	36	113.29	25.7	33	76.74
34.1	41	167.87	29.8	36	112.26	25.6	33	75.79
34	41	166.8	29.7	36	111.8	25.5	33	75.5
33.9	40	161.85	29.6	36	111	25.4	33	74.9
33.8	40	160.95	29.5	36	110.3	25.3	33	74.39
33.7	40	159.92	29.4	36	109.56	25.2	33	73.79
33.6	40	159	29.3	36	108.8	25.1	33	73.2
33.5	40	158.6	29.2	36	107.7	25	33	72.6
33.4	40	157.8	29.1	36	107.3	24.9	32	69.85
33.3	40	156.88	29	36	106.6	24.8	32	69.18
33.2	40	155.24	28.9	35	102.9	24.7	32	68.74
33.1	40	154.31	28.8	35	102.2	24.6	32	68.19
33	40	153.37	28.7	35	101.5	24.5	32	67.59
32.9	39	148.64	28.6	35	100.8	24.4	32	67.04
32.8	39	147.9	28.5	35	100.1	24.3	32	66.5
32.7	39	146.78	28.4	35	99.4	24.2	32	65.95
32.6	39	145.94	28.3	35	98.7	24.1	32	65.4
32.5	39	145	28.2	35	98	24	32	64.9
32.4	39	144.15	28.1	35	97.3	23.9	31	62.3
32.3	39	143.26	28	35	96.6	23.8	31	61.82
32.2	39	142.39	27.9	35	96.2	23.7	31	61.3
32.1	39	141.5	27.8	35	95.24	23.6	31	60.79
32	39	140.6	27.7	35	94.5	23.5	31	60.27
31.9	38	136.15	27.6	35	93.87	23.4	31	59.76
31.8	38	135.3	27.5	35	93.5	23.3	31	59.72
31.7	38	134.4	27.4	35	92.64	23.2	31	58.71
31.6	38	133.6	27.3	35	91.8	23.1	31	58.2
31.5	38	132.75	27.2	35	91.17	23	31	57.7
31.4	37	128.4	27.1	35	90.5	22.9	30	55.36
31.3	37	127.6	27	35	89.84	22.8	30	54.9
31.2	37	126.8				22.7	30	54.4

TABLE OF GAUGING, CONTINUED.

Mean diameter.	Length.	Gallons.
22.6	30	53.95
22.5	30	53.47
22.4	30	53
22.3	30	52.5
22.2	30	52.1
22.1	30	51.5
22	30	51.1
21.9	29	48.94
21.8	29	48.52
21.7	29	48.2
21.6	29	47.6
21.5	29	47.2
21.4	29	46.7
21.3	29	46.3
21.2	29	46.89
21.1	29	45.46
21	29	45
20.9	28	43.6
20.8	28	42.65
20.7	28	42.24
20.6	28	41.83
20.5	28	41.4
20.4	28	41
20.3	28	40.62
20.2	28	40.2
20.1	28	39.83
20	28	39.46
Barrels.		
20.9	29	44.6
20.8	29	44.17
20.7	29	43.75
20.6	29	43.4
20.5	29	42.9
20.4	29	42.49
20.3	29	42.07
20.2	29	41.66
20.1	29	41.25
20	29	40.8
19.9	28	39
19.8	28	38.6
19.7	28	38.26
19.6	28	37.87
19.5	28	37.48

Mean diameter.	Length.	Gallons.
19.4	28	37.1
19.3	28	36.72
19.2	28	36.34
19.1	28	35.26
19	28	35.23
18.9	27	33.96
18.8	27	33.6
18.7	27	33.2
18.6	27	32.82
18.5	27	32.5
18.4	27	32.17
18.3	27	31.8
18.2	27	31.4
18.1	27	31.14
18	27	30.8
17.9	26	29.37
17.8	26	28.98
17.7	26	28.67
17.6	26	28.35
17.5	26	28.03
17.4	26	27.7
17.3	26	27.4
17.2	26	27.38
17.1	26	26.76
17	26	26.4
16.9	25	25.14
16.8	25	24.84
16.7	25	24.5
16.6	25	24.2
16.5	25	23.95
16.4	25	23.67
16.3	25	23.39
16.2	25	23
16.1	25	22.8
16	25	22.5
15.9	24	21.36
15.8	24	21.09
15.7	24	20.87
15.6	24	20.56
15.5	24	20.2
15.4	24	20
15.3	24	19.78
15.2	24	19.52

Mean diameter.	Length.	Gallons.
15.1	24	19.26
15	24	19
14.9	23	17.97
14.8	23	17.73
14.7	23	17.5
14.6	23	17.26
14.5	23	17
14.4	23	17.13
14.3	23	16.54
14.2	23	16.32
14.1	23	16.1
14	23	15.87
13.9	22	14.96
13.8	22	14.75
13.7	22	14.53
13.6	22	14.32
13.5	22	14.11
13.4	22	13.9
13.3	22	13.77
13.2	22	13.49
13.1	22	13.29
13	22	13.28
12.9	21	12.3
12.8	21	12.11
12.7	21	11.92
12.6	21	11.70
12.5	21	11.55
12.4	21	11.36
12.3	21	11.18
12.2	21	11
12.1	21	10.82
12	21	10.6
11.9	20	9.97
11.8	20	9.8
11.7	20	9.64
11.6	20	9.44
11.5	20	9.31
11.4	20	9.15
11.3	20	8.99
11.2	20	8.86
11.1	20	8.67
11	20	8.52

TABLE OF GAUGING, CONTINUED.

Mean diameter.	Length.	Gallons.	Mean diameter.	Length.	Gallons.	Mean diameter.	Length.	Gallons.
10.9	19	7.94	9.9	18	6.21	8.9	17	4.74
10.8	19	7.8	9.8	18	6.08	8.8	17	4.63
10.7	19	7.65	9.7	18	5.95	8.7	17	4.53
10.6	19	7.51	9.6	18	5.84	8.6	17	4.4
10.5	19	7.37	9.5	18	5.72	8.5	17	4.32
10.4	19	7.23	9.4	18	5.6	8.4	17	4.22
10.3	19	7.9	9.3	18	5.47	8.3	17	4.12
10.2	19	6.96	9.2	18	5.35	8.2	17	4.02
10.1	19	6.82	9.1	18	5.24	8.1	17	3.82
10	19	6.69	9	18	5.13	8	17	3.8

JOHN Q. ADAMS'S REPORT.

I shall here introduce quotations from the report of Mr. Adams, Secretary of State, on weights and measures, ordered by the Senate of March 3, 1817, and presented February 22, 1821. The report in full would occupy too much space, or I should be happy to present it; but I shall give enough to show that he was in need of the measure of the circle to complete a useful report. Had Mr. Adams and Mr. Jefferson known the true measure of the circle, their labor to attain their object might be confined to personal judgment altogether, independent of former instructions. Mr. Adams says, —

When weights and measures present themselves to the contemplation of the legislator, and call for the interposition of law, the first and most prominent idea which occurs to him is that of uniformity; his first object is to embody them into a system, and his first wish to reduce them to one universal common standard. His purposes are, uniformity, permanency, universality, — one standard to be the same for all persons and all purposes, and to continue the same forever. These purposes, however, require powers which no legislator has hitherto been found to possess. The power of the legislator is limited by the extent of his territories and the numbers of his people. His principles of universality, therefore, cannot be made by the mere agency of his power to extend beyond the inhabitants of his own possessions. The power of the legislator is limited over time; he is liable to change his own purposes; he is not infallible; he is not immortal; his successor accedes to his power, with different views, different opinions, and perhaps different principles. The legislator has no power over the properties of matter; he cannot give a new constitution to nature; he cannot repeal her law of universal mutability; he cannot square the circle; he cannot reduce extension and gravity to one common measure; he cannot divide or multiply the parts of the surface, the cube or the sphere, by the uniform and exclusive number 10. The power of the legislator is limited over the will and actions of his subjects. His conflict with them is desperate, when he counteracts their settled habits, their established usages, their domestic and individual economy, their ignorance, their prejudices, and their wants; all which are unavoidable in the attempt radically to change or to originate a totally new system of weights and measures.

In the law given on Sinai, not of a human legislator, but of God, there are two precepts respecting weights and measures; first, Ye shall do no unrighteousness in judgment in mete-yard, measure of lengths, in weights, or in measure of capacity. Just balance, just weight, a just ephah, and a just hin shall ye have: thou shalt not have divers of weights, great and small, but thou shalt have perfect and just weights and measures with all.

Among the nations of modern Europe there are two, who by their genius, their learning, their industry, and their ardent and successful cultivation of the arts and sciences, are distinguished not less than the Hebrews, from whom they have derived many of their civil and political institutions. From these two nations the inhabitants of the United States are chiefly descendants; and from one of them we have all our existing weights and measures. Both of them, for a series of ages, have been engaged in the pursuit of a uniform system of weights and measures. The legislators and all exertions by man have been extended to effect this, but without success, for the want of a perfect measure of the circle.

It appears that a reformation is in agitation in Spain, to correct weights and measures. It is now under the consideration of the courts in that kingdom; and, as weights and measures are the necessary and universal instruments of commerce, no change can be effected in the system of any one nation without affecting all those with whom there are any relations of trade and commerce. The results of this inquiry in Spain are not known. France and England are the only two nations of modern Europe that have taken much interest in the organization of a new system, or attempted a reform for the avowed purposes of uniformity. The proceedings with these two nations have been numerous, elaborate, persevering, and in France especially comprehensive, profound, and systematic. In both the phenomenon is still exhibited, that, after many centuries of study, of invention of laws and of penalties, almost every village in the country is in the habitual use of different weights and measures, which diversity is infinitely multiplied by the fact that in each country, although the quantities of the weights and measures are thus different, their denominations are few in number, and the same names, as foot, pound, ounce, bushel, pint, &c., are applied in different places, and often in the same place, to quantities altogether diverse.

In England, from the earliest records of parliamentary history, the statute books are filled with ineffectual attempts of the legislator to establish uniformity of the origin of their weights and measures. The very words of a law of William the Conqueror are cited by modern writers on the English weights and measures. Their import is, "We ordain and command that the weights and measures be as established and commanded by our worthy predecessors." One of the principal objects of the great charter was the establishment of uniformity of weights and measures, not of identity, but of proportion, in order to shut the door of dishonesty against the evasions of good rules and law.

The object of whole statutes was to guard against the fraud and oppression that were used with the people, for the want of a perfect measure. It has been remarked by all historians that the taking of kernels of wheat as a standard of measure is inconsistent with sound philosophy, as the variations of climate, the seasons, and the grains in the same field, are different.

From the year 1757 to 1764, in the years 1789 and 1790, and from the year 1814 to the present time, the British Parliament has at three successive periods instituted inquiries into the condition of their own weights and measures, with a view to the reformation of the system, and to the introduction and establishment of greater uniformity. These inquiries have been pursued with ardor and perseverance, assisted by the skill of their eminent artists, by the learning of their distinguished philosophers, and by the contemporaneous admirable exertions, in the same cause of uniformity, of their neighboring and rival nations.

Nor have the people or the Congress of the United States been regardless of the subject, since our separation from the British empire. In their first confederation, the associated states, and in their present national constitution, the people, that is, on the only two occasions upon which the collective voice of this whole Union, in its constituent character, has spoken, the power of fixing the standard of weights and measures throughout the United States has been committed to Congress. Elaborate reports, one from a committee of the Senate, in 1793, another from the House of Representatives, at a recent period, have since contributed to shed further light upon the subject; and the call of both houses, to which the report is the tardy but yet too early answer, has manifested a solicitude for the improvement of the existing system equally earnest and persevering with that of the British Parliament, though not marked with the bold and magnificent characters of the concurrent labors of France.

To despair of human improvements is not more congenial to the judgment of sound philosophy than to the temper of brotherly kindness. Uniformity of weights and measures is and has been for ages the common, earnest, and anxious pursuit of France, of Great Britain, and, since their independent existence, of the United States. To the attainment of one object, common to them all, they have been proceeding by different means and different ultimate ends. A single and universal system can be finally established only by a general convention, where all the world shall be parties, and by one continual exertion to effect a system of uniformity of weights and measures that shall be meet and just to all.

New Hampshire and Vermont.

These two states have modelled their weights and measures from Massachusetts. The first act of New Hampshire to that effect was of the 13th of May, 1718; and the last, December 15th, 1797. The statutes of Vermont have established the troy weight by law; but each and every state differs in its measures, no two being alike.

Rhode Island

Has no statute on the subject. Her standards are taken from Massachusetts, and yet differ in all weights and measures.

Connecticut.

In the laws and standards of this state there are peculiarities deserving of remark. Their statute of 1800 provides that the half bushel shall contain 1099 cubic inches. This varies from some other states 29 cubic inches in half a bushel. The wine gallon shall be 282 cubic inches, beer measure shall be 224 inches.

New York.

New York was originally the seat of a colony from the Netherlands, the settlers of which doubtless brought with them the weights and measures of their country. Towards the close of the seventeenth century it fell into the hands of the English;

and on the 19th of June, 1703, an act of the colonial legislature established all the English weights and measures, according to the standards in the exchequer. This act was drawn with great care, and evidently with the purpose of embracing all the provisions of the then existing English statute regulating weights, and measures, and casks, particularly those of 1266, 1304, 1439, and 1496, without being aware of the utter incompatibility of those statutes with one another.

Instead, however, of adopting in terms the London assize of casks, from the tun of 252 gallons downward, this act prescribes in inches the length and head diameter of the various casks, and, by a very remarkable peculiarity, changes the names of all dry casks.

It directs that the hogshead shall be 40 inches long, 33 inches in the bilge, and 27 in the head; the tierce, 36, 27, 23; the barrel, 30, 25, 22; half barrel, 25, 20, 16; quarter barrel, 20, 16, 13. But it adds that tight barrels shall contain 31½ gallons, wine measure, or with half a gallon more or less, and all other casks in proportion. This last provision adopted the whole London assize for tight casks. But the dimensions prescribed for the hogshead give the cask about 126 gallons, which in the London assize made the butt or pipe; and thus the New York tierce was of 80 gallons, which constituted the real contents of the London puncheon; the New York barrel was of 60 gallons, answering to the London hogshead; and the New York half barrel of 30 gallons to the London barrel. On the 10th of April, 1784, the Legislature of New York passed an act to ascertain weights and measures within the state. It declares the standard of weights and measures which were in the custody of William Hardinbrook, public sealer and marker in the city and county of New York at the time of independence, which were according to the standard of the exchequer, to be standards throughout the state. William Hardinbrook was directed to deliver them to the clerk of the city and county of New York, and to make oath that they were the same that he had received of the court of the exchequer. By an act of the 7th of March, 1788, the standard weight of wheat was 60 pounds net to the bushel. On the 24th of March, 1809, an act was passed relative to a standard of long measure, and in 1813 another act.

New Jersey.

In New Jersey, which was originally a part of the Dutch settlement, the English standard of weights and measures was established at a later period than in New York. An act of the colonial legislature of the 13th of August, 1725, recites in the preamble that nothing is more agreeable to common justice and equity than that throughout the province there should be one just weight and balance, one true and perfect standard of measures, for want whereof experience has shown that many frauds and deceits had happened; for remedy of which it established, in the first section, an assize cask for the packing of beef and pork, since altered. The second declares that there shall be one just and equal balance, one certain standard for weights, that is to say, for avoirdupois and troy weight, one standard measure for the half bushel, peck, and half peck, and one for liquid measure; all which shall be according to the standard of Great Britain in the exchequer.

This appears to be the standard of New Jersey, but there is a general variation in all their weights and measures, which differ in size and weight from many others. They have had many changes in relation to the assize of barrels.

Pennsylvania.

In the year 1700, two laws relating to weights and measures were enacted by the colonial Legislature, directing that a brass standard from the exchequer of England should be the standard. The second act not only adopted the London assize of cask, but required that all tight casks for beer, ale, cider, pork, beef, oil, and all such commodities, should be made of good, sound, well-seasoned white oak timber, and contain, — the puncheon, 84 gallons, the hogshead, 63 gallons, the tierce, 42, the barrel, 31½, the half barrel 16 gallons; wine measure to be according to the practice of the neighboring colonies.

As in all other states, there is a continual variation in their weights and measures, for the want of a perfect measure, the state authorizing the innkeepers to sell beer by wine measure.

Delaware.

In 1705 an act for regulating weights and measures was passed, directing that each county should obtain standard brass weights and measures, according to the queen's standard for the exchequer. The half bushel was to be taken from Philadelphia. It authorizes the standard of England. The same variation exists in the measures of this state from all others.

Maryland.

This state has had all kinds of legislative acts upon weights and measures, varying, in all kinds of measurement, from most other states. The peculiar constructions upon weights and measures in this state are numerous, and wonderful in the extreme.

Virginia.

Among the earliest records of the General Assembly of the colony of Virginia is an order of the 5th of March, 1623–4, that there should be no weights and measures used but such as should be sealed by officers appointed for that purpose. In this state they have had all kinds of measure, varying in every form and manner that ingenuity could devise.

North Carolina.

The only law of this state relating to weights and measures, a knowledge of which has been obtained, was enacted prior to the American revolution, during the administration of Governor Gabriel Johnson, and is yet in force. It prohibits the use in trade, by all the inhabitants or traders within the province, of any weights and measures other than are made and used according to the standard in the English exchequer. It charges the justices of the county courts to provide, at the expense of each county, the standards. Some of their measures are singular; oil, beer, wine, bushels, pecks, are the same; and all measurement is after this manner, varying with themselves, and every body, and every thing.

South Carolina.

By an act of the 12th of April, 1768, the public treasury was required to procure, of brass, or some other metal, one weight of 50 pounds, one of 25, one of 14, two of 6 pounds, two of 4 pounds, two of 2 pounds, and one of 1 pound, avoirdupois weight, according to the standard of London, and one bushel, one half bushel, one peck, and one half peck, according to the standard of London. These were to be stamped or marked in figures denominating their weights and measures, and were declared to be the standard by which all others in the province were to be regulated.

One continual variation exists in this state, as in all others; dry and wet measures are used as the same.

Georgia.

An act of the Legislature of the 10th of December, 1803, declares the standard of weights and measures established by the corporation of the cities of Savannah and Augusta to be the fixed standard of weights and measures within the state; and that all persons buying or selling shall use that standard until the Congress of the United States shall make provision on that subject. It directs the justices of the inferior courts in the counties to obtain standards conformable to those of the corporation of one of those cities.

An ordinance of the city council of Augusta directs that all weights for weighing any articles of produce or merchandise shall be of the avoirdupois standard weight; and all measures of liquor, whether of wine or ardent spirits, of the wine measure standard; and all measures for grain, salt, or other articles usually sold by the bushel, of the dry or Winchester measure standard. And it prohibits the use of any other than brass or iron weights, thus regulated, or weights of any other description than those of 50, 25, 14, 7, 4, 2, 1, ½, and ¼ pounds, 2 ounces, 1 ounce, and downward.

Kentucky.

An act of the Legislature, of the 11th of December, 1798, stating in its preamble that Congress is empowered by the federal constitution to fix the standard of weights and measures, and that they had not passed any law for that purpose, recognizes as thereby remaining in force within that commonwealth the act of the General Assembly of Virginia of the year 1734. It therefore authorizes and directs the governor to procure one set of the weights and measures specified by the Virginia act of 1734, with measures of the length of one foot and one yard, and declares that the bushel, of dry measure, shall contain $2150\frac{2}{3}$ solid inches, and the gallon, of wine measure, 231 inches.

Here is a difference in the bushel of 106 inches from some states, and in the wine measure of 10 inches in one gallon from some other states under this government.

Tennessee.

From a communication received from the governor of the State of Tennessee, it appears that there is in that State no standard of weights and measures fixed by the Legislature.

Ohio.

The only act of the Legislature of the State of Ohio on this subject is of the 22d of January, 1811. It directs the county commissioners of each county in the State to cause to be made one half bushel measure, to contain $1075\frac{2}{10}$ inches, solid, which is to be kept in the county seat, and be called the standard.

Louisiana.

Before the accession of Louisiana to the Union of these States, the weights and measures used in the provinces were those of France, of the old standard of Paris. An account of these, and of the present state of the weights and measures in the State of Louisiana, is submitted in the appendix to this report. This State uses for a barrel 330 cubic inches less than some other States, all under one government.

Indiana.

An act of the territorial Legislature of the 17th of September, 1807, authorized the courts of common pleas of the respective counties in the territory, whenever they might think it necessary, to procure a set of measures and weights for the use of the country, — one measure of 1 foot, or 12 inches, English measure, so called, one measure of 3 feet, or 36 inches, English measure, and half a bushel, of dry measure, to contain $1075\frac{1}{5}$ solid inches, and one gallon measure, to contain 231 inches.

This State differs in the dry measure from some of the States 53 inches in the bushel; other measures in the same proportion, or more.

Mississippi.

An act of the territorial Legislature, of the 4th of February, 1807, directed the treasurer to procure a set of the large avoirdupois weights, according to the standard of the United States, if one were established, but if there were none such, according to the standard of London, with proper scales for weights, together with measures of the foot and yard, dry measure of capacity, and liquid wine measure.

This standard is to remain until the United States fixes a standard. A barrel of flour contains 196 pounds; pork and beef, 200 pounds to the barrel.

Illinois.

The territorial act of the 17th of September, 1807, was passed while the State of Illinois formed a part of the Indiana territory; but by the act of the Legislature of this State, regulating weights and measures, of the 22d of March, 1819, the county commissioners of each county in the State were required to procure, at the expense of the county, one foot, one yard, English measure, a gallon liquid or wine measure, to contain 231 cubic inches, a corresponding quart, pint, and gill measure, of some proper metal, a half bushel, dry measure, to contain eighteen quarts, one pint, and one gill, wine measure, or $1075\frac{2}{10}$ cubic inches, and a gallon, dry measure, to contain one fourth part of half a bushel; these measures to be copper or brass. Also, weights, one pound, $\frac{1}{2}$ pound, $\frac{1}{8}$, and $\frac{1}{16}$, made of brass, the integer of which to be denominated one pound, avoirdupois, and to equal in weight 7020 grains, troy or gold weight.

The most remarkable peculiarity of this act is, its departure from the English standard weights by fixing the avoirdupois pound at 7020 instead of 7000 grains troy.

Alabama.

This State having formed a part of the Mississippi territory previous to the admission of the State of Mississippi into the Union, in 1817, the acts of that territory of the 4th of February, 1807, and the 23d of December, 1815, embraced this section of the territory. No act of the State Legislature of Alabama on this subject is known to have been passed.

Missouri.

The territorial Legislature, by an act of the 28th of July, 1813, directed the several courts of common pleas within the territory to provide, at the expense of the respective counties, one foot, one yard, English measure, one half bushel, to contain $1075\frac{1}{5}$ solid inches, for dry measure, one gallon, to contain 231 inches, and smaller liquid measures in proportion, to be of any wood or any metal the court think proper; also, one set of avoirdupois weights. The use or keeping, to buy or sell, of weights or measures not corresponding with these standards, after due notice, was prohibited, under penalties, by the same act, but with a provision that all contracts or obligations made previous to the taking effect of the act should be settled, paid, and executed, agreeably to the weights and measures.

District of Columbia.

By the act of Congress of the 27th of February, 1801, concerning the District of Columbia, the laws of the State of Virginia, as they then existed, were continued in force in the part of the district which had been ceded by that State, and the laws of Maryland in the part of the district ceded by Maryland.

The act to incorporate the inhabitants of the city of Washington, of the 3d of May, 1802, authorizes the corporation to provide for the safe keeping of the standard of weights and measures fixed by Congress, and for the regulation of all measures used in the city. The supplementary act of the 24th of February, 1804, gives the city council power to establish and regulate the inspection of flour, tobacco, and salted provisions, and the gauging of casks and liquor.

There have been in this District numerous changes and alterations.

As preliminary remarks in reference to that part of the resolutions of both houses which requires the opinion of the Secretary of State with regard to the measure which it may be proper for Congress to adopt in relation to weights and measures, I will state what might be done in Congress: in one line of the constitution to adopt the French standard. To fix the standard appears to be an operation entirely distinct from changing the denominations and proportions already existing, and established by the laws, or immemorial usage. In Europe, every historical research presents a fruitless struggle to effect a uniform system of weights and measures; less exertion, perhaps, exists in the

United States for the weights and measures than elsewhere. The United States adopted their measures and weights from England, or from usage; and no general exertions have taken place, like those of the old country. The wine gallon of 231 inches, and the beer gallon of 282 inches, have been known by usage, and the Winchester bushel of 2150.42 inches formed the general standard. The various multipliers of the yard, ell, perch, pole, furlong, acre, and mile were recognized by law.

The experience of the French nation under the new system has already proved that neither the immutable standard from the circumference of the globe, nor the isochronous vibration of the pendulum, nor the gravity of distilled water at its maximum of density, nor the decimation of weights, measures, moneys, and coins, nor the unity of weights and measures of capacity, nor yet all these together, are the only ingredients of practical uniformity for a system of weights and measures. It has proved that gravity and extension will not walk together with the same staff; that neither the square, nor the cube, nor the circle, nor the sphere, nor the gravitation of the earth, nor the harmonies of the heavens, will gratify the pleasure, or, to indulge the indolence of man, be restricted to computation by decimal numbers alone.

A perfect measure of the circle may be considered as entering into the economical arrangements and daily concerns of every family. It is necessary to every occupation of human industry, to the distribution and security of every species of property, to every transaction of trade and commerce, to the labors of the husbandman, to the ingenuity of the artificer, to the studies of the philosopher, to the researches of the antiquarian, to the navigation of the mariner, and the marches of the soldier; to all the exchanges of peace, and all the operations of war. It is among the first researches of literature; the establishment of its truth is among the first elements of education. This knowledge is riveted in the memory by the habitual application of it to the employments of men, through life. Every individual, or, at least, every family, has the weights and measures used in the vicinity, and recognized by the custom of the place. To change all this at once is to affect the well being of every man, woman, and child in the community. It enters every house, it cripples every hand.

Weights and measures, and the final establishment of a system for them, with a view to the utmost practical extent of uniformity, are at this moment under the deliberate consideration of four populous and commercial nations, — Great Britain, France, Spain, and the United States. The interest is common to them all; the object of uniformity is the same to all. Could they agree upon one result, the advantages of that agreement would be great to each of them, and still greater in all their intercourse with one another. It is therefore respectfully proposed, as the foundation of proceedings necessary for securing ultimately to the United States a system of weights and measures which shall be common to all civilized nations, that the President of the United States be requested to communicate, through the ministers of the United States in France, Spain, and Great Britain, with the governments of those nations, upon the subject of weights and measures, with reference to the principle of uniformity, as applicable to them.

The Pendulum. — It is proposed to discard all considerations of the pendulum, as the theory of its vibration, however interesting in itself, is believed to be, since the definitive determination of the metre, useless with reference to any system of weights and measures.

The most prominent numbers made use of in the Bible are 3, 6, 9, and 12. The number 6 prefigures time, and measures the circle, with the hexagon, the hexagon being equal in all its parts. These numbers — 3, 6, 9, and 12 — are used in the performance of all principal important things where numbers are used — 12 apostles, 12 signs in the zodiac, and 3 persons in the Godhead, Father, Son, and Holy Ghost.

MEASURE BY CUSTOM HOUSES IN THE UNITED STATES.

The following are the contents of the different dry measures, by admeasurement, and by the weights of water they contain, as obtained at the different custom houses by the investigation directed by the late President John Q. Adams, reduced to the bushel derived from each: —

Names of the Custom Houses	Cubic Inches.	Avoirdupois Weight Lbs.	oz.	dwt.
Bath, Me.,	1925.00	74	2	0
Belfast,	2063.76	76	0	0
Frenchman's Bay,	2216.70	84	7	8
Kennebunk,	2203.32	78	0	0
Machias,		75	4	0
Lubec,	2158.33	78	14	0
Portland,				
Falmouth,				
Saco,	2215.80	80	0	0
Wiscasset,				
Portsmouth, N. H.,	2153.74	77	12	0
Boston,	2211.06	78	4	0
Newburyport,	2150.52			
Gloucester,	2150.40	78	8	0
Dighton,	2062.78	75	13	0
New Bedford,	2155.12	77	13	8
Barnstable,	2153.82	77	0	0
Edgartown,				
Nantucket,				
Providence, R. I.,	2194.50	78	8	0
Bristol,	2155.13	78	0	0
Newport,	2160.18	77	14	0
New London, Ct.,	2222.06	78	10	0
Fairfield,	2249.86	79	0	0
New York, mean,	2152.56	78	13	4
Rochester, N. Y.,	2202.58			
Philadelphia,	2186.20	78	12	0

Names of the Custom Houses.	Cubic Inches.	Avoirdupois Weight. Lbs.	oz.	dwt.
Wilmington, Del.,	2192.20	77	4	12
Baltimore, Md.,	2150.42	77	8	0
Oxford, Md.,	2060.94	80	0	0
Washington, D. C.,	2117.20	76	7	10
Georgetown, D. C.,	2152.60	77	14	2
Alexandria,	2118.80	77	11	0
Cherrystone,	2225.48	83	4	0
Norfolk, Va.,	2127.24	78	0	0
Petersburg,	2147.08	78	0	0
Richmond,	2112.60	78	8	0
Camden, N. C.,	2152.20	79	8	0
Edenton,	2160.78	77	6	0
Newburn,	2115.60	87	8	0
Ocracoke,	2153.10	76	0	0
Plymouth,	2358.58	77	0	0
Washington, N. C.,	2128.02	72	12	0
Charleston, S. C.,	2172.03	77	12	12
Savannah, Ga.,	2013.32	76	0	0
St. Mary's, Ga.,	2019.34	78	4	0
New Orleans,	2162.02	77	11	7

In the United States, their coins, both gold and silver, are legal tender for payment, to any amount; but in England, silver coin is a legal tender for payment only to an amount not exceeding 40 shillings; and by the restrictions of each payment by the bank, the only actual currency, the only material, in which an American merchant having a debt due to him in England can obtain payment is Bank of England paper; so that at this time the material of exchange between the United States and England is, on the side of the United States, gold or silver, and on the side of Great Britain, bank paper. Suppose an American merchant has a debt due him in England, which is remitted to him in gold bullion, or coin of the English standard, say £10,000. He receives of pure gold 196 pounds, 2 ounces, 3 pennyweights, 22 grains, for which, when coined at the mint of the United States, he receives 45,657 dollars, 20 cents. The pound sterling therefore yields him 4 dollars, 56.572 cents, which is the value of a pound sterling if the par of exchange be estimated in gold, according to the standard of purity common to both countries. If the payment should be made in silver bullion, at 66 shillings the pound, troy weight, according to the present English standard of silver coinage, he would receive only 43,489 dollars and 43 cents; and the pound sterling would only net him 4 dollars, 34.8943 cents. The pound sterling, estimated in gold, is worth 4 dollars, 56.5720 cents; in silver it is worth 4 dollars, 34.8943 cents; making a difference of 21.6777 cents, half of which, 10.8388, added to $4.348943, and deducted from $4.565720, makes what is called the medium par exchange, $4.457331.

One pound, troy weight, of uncoined gold, or foreign gold coin, eleven parts fine and one part alloy, is $209.77; one pound of silver, eleven parts fine and one part alloy, is $13.77. In April, 1816, an act was passed, regulating the currency, within the United States, of the gold coins of Great Britain, France, Portugal, and Spain, the crowns of France, and 5-franc pieces. By this act, gold coin of Great Britain and Portugal weighing 27 grains equal 100 cents, or 88⅘ cents per dwt.; France, 27½ grains equal 87¼ cents; Spain, 28½ grains equal 84 cents; crowns of France weighing 449 grains equal 110 cents, or $1.17 per ounce; 5-franc pieces weighing 386 grains, 93.3, equal $1.16.

One pound, troy weight, of standard gold in England contains 5280 grains of pure gold. It is coined into £46 14 s. 11.214 d. Then 11.214 : 5280 : : 240 : 113.0014 grains of pure gold in a pound sterling.

In the United States, 24.75 grains of pure gold are coined into a dollar, or 247.5 grains to one eagle. Thus 24.75 : 1 : : 113.0014 : $4.56572 to 1 pound. Thus the pound sterling in gold is worth $4.56572; and as 5280 : 11.214 : : 24.75 : 52.5656 dollars in English gold, 4 s., 4.5656 pounds sterling in gold, $4.56572.

One pound, troy weight, of standard silver, in England, contains 5328 grains of pure silver, and is coined into 66 shillings, or 792 pence. The dollar of the United States contains 371.25 grains of pure silver. Then 5328 : 792 : : 371.25 : 55.1858 dollars in English silver, 4 s. 7.1858 d.; 792 : 5328 : : 240 : 1614.545 grains of pure silver in a pound; 371.25 : 1614.545 : : £1 4.348943 s. sterling in $4.348943, silver; medius par dollar, 4 s. 5.8757 pence. £1 sterling in gold, $4.565720 — 10.8388 = $4.457331, med. par; £1 sterling in silver, $4.348943 +.

We are told that 1800 circumference will not work in proportion as 6 to 19, as other numbers.

If 19 : 6 : : 1800
6
———
19) 10800 ($568\frac{8}{19}$
95
——
130
114
——
160
152
——
8

I add to the $568\frac{8}{19}$ the $31\frac{11}{19}$.
$31\frac{11}{19}$
——
600
This is the diameter of 1900.

If 18 : 1 : : $568\frac{8}{19}$
1
———
18) $568\frac{8}{19}$ ($31\frac{11}{19}$
54
——
28
$18\frac{8}{19}$
—
10
Multiply by 19
—
90
10
——
$198\frac{11}{19}$
18
——
18
18
—
0

If 600 : 1900 : : $568\frac{8}{19}$, answer 1800.

I now find the area of $568\frac{8}{19}$.

```
          568.421
          568.421
          -------
          568421
         1136842
        2273684
       4547368
      3410426
     2842105
     -------------
     323101.433241   Take 7/9 of this.
                 7
     -------------
  9 ) 2261710.032631
     -------------
```

I divide this by 28.

```
28 ) 251301.113631 ( 8975.09772
     224
     ---
      273
      252
      ---
       210
       196
       ---
        141
        140
        ---
         111
          84
         ---
          273
          252
          ---
           216
           196
           ---
            203
            196
            ---
             71, circum.]
             56
             --
             15, rem.
```

```
     251301.113631
       8975.039772
     -------------
 3 ) 242326.073859
     -------------
 √ 80775.357953 ( 284.210
   4                This is the
   --                  radius.
48) 407
    384
    ----
564) 2375
     2256
     -----
5682) 11935
      11364
      ------
56841) 57179
       56841
       ------
       33853, remainder.
```

```
284.210, radius.
      2
-------
568.420, diameter.
```

I now take the diameter, and find its circumference. 284 × 2 = 568, diameter.

```
     568
     9.5
    ----
    2840
   5112
   -----
3 ) 53960
   -----
   1798.6
        2
   ------
   1800
```

I add the 2 for the $\frac{8}{19}$, and call the remainders equal to the 2 remaining.

In the first place I show that 6 to 19 brings the diameter of a circle of 1800 to be $568\frac{8}{19}$; then I show that $\frac{1}{19}$ is $31\frac{11}{19}$; now add this to the $568\frac{8}{19}$, which gives 600. This 600 is the diameter of 1900; then I show that if 600 gives 1900, $568\frac{8}{19}$ will give 1800; then I find the area of the $568\frac{8}{19}$; then I find the circle that bounds the area, which is 1800.

$$
\begin{array}{r}
568.421\frac{1}{19} \\
9.5 \\
\hline
2842105 \\
51157895 \\
\hline
3\,)\,54000000 \\
\hline
1800.0000
\end{array}
$$

The rule to find the circle is, to divide by 28, subtract the quotient, divide by 3, get the square root, and that will be the radius of the circle that bounds the figures.

On the following page I have given the working of 1500 circumference. I have worked it as far as it is possible, because it begins to repeat, as you may see. It begins with 90, and comes to 90 again. It is what is called a surd; and it is said its root can never be perfected. I say it cannot be perfected by decimals, but by a vulgar fraction it can, as 4 pence are just as perfectly $\frac{1}{3}$ of a shilling, as $\frac{5}{10}$ are equal to $\frac{1}{2}$, except that one is a vulgar fraction, and the other a decimal.

First, what is the diameter?—

If 19 : 6 : : 1500
6

19) 9000 (473.6842105263157899$\frac{9}{19}$
76
140
133
70
57
130
114
160
152
80
76
40
38
20
19
100
95
50
38
120
114
60
57
30
19
110
95
150
133
170
152
180
171
90

I find the diameter more in whole numbers:—

If 19 : 6 : : 1500
6

19) 9000 (473$\frac{13}{19}$
76
140
133
70
57
13

```
                      473
                       13
                     ----
                     1419
                     473
                     ----
                19 ) 6149 ( 323 × 2 = 647, plus.
                     57
                     ----
       473            44
       473            38
      -----          ----
      1419            69
     3311             57
    1892              --
       447, added.    12
    ------             8, for the corner.
    224376            --
         7            20
    ------
9 ) 1570632 ( 174914⅔+    This is the area of 1500
    9                          circumference.
    --
    67
    63
    --
     40
     36
     --
      46
      45
      --
       13
        9
       --
        42
        36
        --
         6  This remainder is equal to ⅔.
```

To find the measure of a globe or sphere, or, rather, if a certain number of miles, feet, inches, or whatever you may please to call it, be made to compose a globe, what the diameter of that globe will be: Suppose 8064 to be the given number; I now divide 8064 by 7, and multiply that quotient by 5, the product of which I add to the given number, thus:—

```
 8064
 5760
─────
13824
```

I now extract the cube root of this number, and the quotient will be the diameter of the globe.

```
7) 8064
   ────
   1152
      5
   ────
   5760   Add this to the given number, 8064.
   ────
√ 13824   I find the cube root of this.
  13824 ( 24 is the cube or diameter of globe.
  8
  ────
  5824, first resolved.
     6, three times the root.
    12  three times the square of the root.
   ────
   126, the first divisor.
   ────
    64, the cube of 4.
   96   square of 3, multiplied by 3 times 2.
  48    three times the square of 2 by 4.
  ────
  5824, subtrahend.
  ────
  0000
```

Proof: $\frac{7}{12}$ of the cube is the globe, and the cube is 13824; therefore, divide the cube by 12, and 7 times the quotient is the globe:—

```
12 ) 13824
     -----
      1152
         7
     -----
      8064, answer.
```

Having examined Messrs. Hodson and Pike's works, I find that in extracting the cube root, having found the divisor, they place one less than it will go, which is right in the sum laid down; but I think they ought to have given reason for so doing, otherwise it is a great puzzle. Now, in the sum on the preceding page it gives just the number of times; but in that laid down by Pike it goes one less, thus:

```
√³ 16194277 ( 253
    8
    ----
    8194, subtrahend.
    ----
     126, divisor.
    ----
     116
    216
    72
    ----
    9576, subtrahend.
    ----
      125
Thus. 150
      60
     ----
     7625, right subtrahend.
```

A GENERAL VIEW OF THE WORK.

I show by the strips of paper, which are 9 inches long, that the vacant corners are the biquadrate of the circle. The biquadrate is the 4th power; it is the $\frac{1}{18}$ of the hexagon, or the $\frac{1}{18}$ of 3 times the diameter added, which completes the measure of the circle.

The area of the dodecagon can be found by the six oblong squares, as laid down in my work; but it does not find the circular part outside of the dodecagon, which is just the square of the biquadrate, and that added will complete the area of any circle, which is the 27th part of the dodecagon, added to the dodecagon makes a 28th part of the whole. This biquadrate is the 4th power of the number 6; but a biquadrate is the 4th power of any number.

In a circle 6 inches in diameter the biquadrate gains 1 inch upon a line; therefore the square of that inch will make just one square inch in the area; whereas a circle 12 inches in diameter gains 2 inches, linear measure, making just 4 square inches in the area, as $2 \times 2 = 4$.

The square of what the biquadrate gains in circumference is what the circle gains in area; so that the 4 inches is just what is contained without the dodecagon, which is the circular part of the circle, and is what makes up the measure of the circle. The biquadrate descending is 1296 parts of one inch; the biquadrate in a 12 inch circle is 2 inches on a line, and the square of that is the area, which is 4.

I show that I find the area of a circle by dividing the circle into 6 angles, and forming an oblong square from one of them, equal to the product of the radius by half the radius, which finds the square inches, and by adding the biquadrate to the square inches, makes up the measure of the circle.

I am asked what evidence I have to prove that the proportion the diameter of a circle has to its circumference is as 6 to 19? I answer, there is no other way to prove that an apple is sour, and why it is so, than by common consent. For example: I will suppose that the diameter of a circle to its circumference is as 6 to 19. I then suppose a circle 12 inches in diameter to be filled with rings $\frac{1}{6}$ of an inch wide, and 36 in number. I suppose the outside ring to be on its outside 38 inches. I find that all these rings measure 114 inches, if I measure them in straight strips, and square at the ends; the average length of the rings on a straight line is 19 inches.

Now, to form a circle of these rings, I take $\frac{2}{6}$ of an inch from the inside of the ring, which goes to the length to form the circle. I then find that these 36 rings, $\frac{1}{6}$ of an inch wide, measure 112 inches; and as my diameter is 12 inches, and my longest ring is allowed to be 38 inches on the outside, I find the proportion is as 6 to 19; as 3 times 12 are 36, and 2 are 38; and 3 times 6 are 18, and 1 are 19; and 4 times 9 are 36, and 2 are 38; and twice 9 are 18, and 1 are 19; and 4 times 9 are 36, and 2 are 38. My proportion of the diameter to the circumference is as 6 to 19, and my ratio is 9.5.

I defy mathematics to prove any other proportion that the diameter of a circle has to its circumference. Why not ask, with the same propriety, what proportion the square has to the square, when it is as 12 to 12; how can I prove that is the proportion? If it is 144 inches, why is it a square?

There are 18 perfect equilateral angles in any circle, great or small, which is 60 degrees, and there is no other triangle that will be equal but 60 degrees. The curve line gains $\frac{1}{18}$ over the straight, on every side of the hexagon; therefore the proportion must be as 6 to 19.

A circle 6 inches in diameter is 28 inches in area; and one 12 inches in diameter is 112 in area. Twice the diameter is 4 times the area, and so in proportion.

Archimedes' measure, as 7 to 22, is just $\frac{1}{4}$ of an inch in 33 too short, almost $\frac{3}{10}$ in the yard, and near $\frac{1}{10}$ in the foot. The biquadrate is equal to $\frac{1}{36}$ of $\frac{1}{36}$; and the sides of the hexagon being divided by 36, the two sides of the biquadrate will equal $\frac{2}{36}$, or $\frac{1}{18}$, on one side of the hexagon, and the 6 sides will gain 2 inches, as 2 added to 36 equal 38. The average measure of the rings is $3\frac{1}{3}$ inches. Every thing in a circle has equal proportions; it is the most perfect thing in nature; every part must have a natural and proportional result. The hexagon gives $\frac{1}{3}$ of $\frac{1}{6}$ for curve, in every part, and $\frac{1}{3}$ of $\frac{1}{6}$ is $\frac{1}{18}$ of itself; so the circle gains $\frac{1}{18}$ of every side of the hexagon; therefore it gains $\frac{1}{18}$ of the whole; so if the 6 sides of the hexagon be

36 inches, the circle is 38, as 7 to 22 is short of correct measure $\frac{1}{8}$ of an inch.

The use of the measure of the circle is, to find the circumference of any circle, great or small, in order to correct all weights and measures, which are wrong, and have been since the world began. No man knows what an inch is, a foot, a yard, a rod, a degree, a peck, a quart, a half bushel, a bushel, a barrel, a hogshead, a pipe, a gallon, or any other measure.

No man can find a perfect level on the earth, for the want of this measure; no man can find the contènts of a steamboat boiler, except by this measure; no man can measure a piece of land, if in a circular form, without a perfect measure of the circle; no man can tell the number of inches in a grindstone, without a perfect quadrature of the circle; no man can gauge a cask of liquor without this measure; no man ever did or ever can extract the square root of the surd number without this measure; no man can give the contents of a globe or sphere without it.

All circular mechanical operations labor under great disadvantages, for the want of the quadrature of the circle. The calculations made by our astronomers of the earth's traverse round the sun is very imperfect, inasmuch as they make it in every 24 hours 1,500,000 miles; whereas correct measure makes it over 1,700,000 miles.

To show the absurdity of the expression from so called great men, such as G. B. Arry, professor royal of England, that the circle is measured near enough for any thing, I will show the difference between the measure of a globe or sphere, taken from books now in use, and the perfect measure. The present measure makes the number of cubic inches to be 904.7808; I make it 1008 by perfect measure. So you see the difference in a globe 12 inches in diameter is 103 inches, and a decimal of 2192.

The present measure makes the convex surface of a globe 12 inches in diameter 452.3804; I make it 361 only. This makes a difference of 91 inches, and 3804 decimal.

Look at this, if measure is near enough. All astronomical observations, which have been attended with great labor and anxiety, and much loss of time, can by this measure be demonstrated with perfection and ease.

The difference between as 7 to 22 and as 6 to 19 is as $\frac{1}{132}$ is to $\frac{1}{133}$. This, as I have before said, makes the foot rule of 12 inches near $\frac{1}{10}$ too short, and the yardstick near $\frac{3}{10}$. Linear measure is .75, square measure 1.52½, cubic measure 2.29 per cent. astray. The proportion the diameter has to the circumference is as 6 to 19; my ratio is 9.5.

Let the cooper take three times the diameter, with one third of the radius, and his head will just fill.

Take a thin strip of tin or brass, the thinner the better, cut a strip 38 inches long, and form a perfect circle, and the diameter will be 12 inches.

To show the difference of area between the square and the circle: Cut a piece of wire 38 inches long, form a perfect circle, and the area will be 112 inches. Cut another piece 38 inches long, and form a square of 9½ inches, and the area contained in the square will be 90¼ inches. Thus the difference of area between the circle and the square will be 21¾ inches.

A gentleman by the name of North, a descendant from Lord North, of England, contemplates building a glass globe, one mile in diameter, in the United States, with the principal places in the world painted within the globe, so that, by the construction of seats within, the people can sit with ease, and see the operation of the earth, and all its principal places. How many square feet of glass would it require? As St. Paul says, in his epistle to the Romans, chapter 11, verse 14, "I do this, if by any means I may provoke any of you to emulation" in measuring the circle. It will take 68,889,600 square feet, making no allowance for deductions.

In a former part of my work I have given a table from John Q. Adams, that people may see the varieties of measures that are made use of, whereas they all ought to be one. These varieties exist only for the want of a perfect measure of the circle, which will be meet and just to all.

I will here give the principal rules that are necessary for practical use, which may be learned by a common scholar in twenty-four hours.

Suppose your circle is 12 inches in diameter; multiply the diameter by 9.5, which is my ratio for finding the circumference; divide the product by 3, which gives the perfect circumference in all cases. Thus:—

```
      12
      9.5
     ———
     108
       6
     ———
  3 ) 114
     ———
      38, circumference.
```

To find the area of the same: Take 3 times the radius by once the radius, which gives the square inches; divide the square inches by 3 to the 4th power or biquadrate; add the 4th power or biquadrate to the square, which gives the perfect area, in all cases. Thus:—

```
      6
      3
     ——
     18                 108
      6                   4
     ——                 ———
 3 ) 108                112, area.
     ——
 3 ) 36
     ——
 3 ) 12
     ——
      4th power, or biquadrate.
```

Or you may take $\frac{7}{9}$ of the square of the diameter, which will give you the perfect area, in all cases. This is more simple and more easy than the above rule. Thus:—

```
        12, diameter.
        12
       ----
       144
         7
      ----
  9 ) 1008
      ----
       112
```

Having the circumference, to find the diameter: Suppose your circumference is 38; divide the circumference by 19, and multiply the quotient by 6, which gives the diameter, in all cases. Thus:—

```
19 ) 38 ( 2
     38   6
     --  --
      0  12, diameter.
```

Having a promiscuous number of figures, to find the circle that bounds them: Divide the figures by 28, subtract the quotient from the sum; divide the remainder by 3; then extract the square root of the quotient, which will be the radius of the circle that bounds the figures; and the square root of any number of figures wrought in this way will be the radius of any circle that bounds the figures. Thus:—

```
28 ) 448 ( 16              448
     28                     16
     ---                   ---
     168              3 )  432
     168                   ---
     ---                   144 ) 12, radius.
       0                   1
                           ---
                     22 )  044
                            44
                           ---
                             0
```

The radius is 12, and twice the radius is the diameter; so the diameter is 24.

I have given the principal rules both in the first and last pages, in order that persons may see the principal practical rules which will enable them to effect all operations of the measure.

Equality and proportion are now demonstrated; space and distance, which never have been known, can now be determined. The difference of the square and the circle can now be told; the area of both is now demonstrated; what the circle gains over the square is illustrated; the difference of the two lines which are called arc and line is known.

What is contained in the dodecagon is now known; a variety of questions which have slept in oblivion can now be mathematically demonstrated, and used for the benefit, happiness, and pleasure of all the learned, and the lovers of science, who may wish to benefit mankind.

In the course of my exertions to measure the circle I have been met with all kinds of objections that ingenuity and art could contrive. I have been told that trigonometry finds the measure of the circle to a hair's breadth.

Now, this is not true. Trigonometry has never been brought to such perfection as to measure a circle. Its use has been to find distance and space, the bounds of which are already stationed and found. Its use is also to find the measure of triangles, and their area, that is, to find three sides; and as it has been used, it has never found but two, in its operation, to measure a circle.

I shall endeavor to show how it can be perfected by trigonometry. I take the number 6, it being the most perfect number that I can find for my use. Now, the hexagon is composed of 6 equal triangles of 60 degrees each; but for convenience I divide them into 36 degrees; and as the triangle gains $\frac{1}{18}$, to make the curve of the circle, I call it 38, two above an equal triangle; and 6 times 38 equal to 38 inches for the circumference of 12 diameter.

Every circle that is twice the diameter of another gains four times as much for area, just the same as the square does. Suppose a square of 3 inches; then 3 inches multiplied by 3 give 9, area; then 6 inches multiplied by 6 give 36, and 4 times 9 are 36; so 12 multiplied by 12 is 144, and 4 times 36 are 144. Now, if the circle does not come in the same proportion, it proves itself wrongly measured; but if it does come in the same proportion, it proves itself right.

Furthermore, if in the first place there is ever so small an error, the error will increase in a fourfold proportion. Thus, if in 3 inches diameter there be a very small error, in 6 inches the error will be 4 times as great, and in 12 inches 16 times, which proves the work, beyond all doubt.

The area of the circle is $\frac{7}{9}$ of the square; so 3 times 3 is equal to 9 square inches; and $\frac{7}{9}$ of that is 7 inches, which is the area of a circle 3 inches in diameter. Now, 6 inches diameter is 4 times 7, equal to 28; and 12 diameter is 4 times 28, which is 112, the area of a circle 12 inches in diameter. And you may carry it as far as you please, which will prove the measure to perfection.

Now, in trigonometry the circumference loses $\frac{1}{57}$ by bringing the curve to a straight line, which makes the diameter in the centre of a triangle of 12 inches 112 instead of 114 sixths of an inch; and the radius being 6 inches, every sixth in the centre makes 1 inch; so 112 inches is the perfect area of a circle 12 inches in diameter. A string that is 38 inches in length in a circle is $\frac{1}{57}$ shorter on a straight line.

I take trigonometry to find the area of one sixth of a circle 12 inches in diameter, and it gives 18.66$\frac{2}{3}$; but by square measure it would be 19 inches, gaining 6 for the curve of the dodecagon; but by trigonometry it makes but 4, which added to 6 times 18, or 108, makes 112 for area; or you may take the rule of three for the preceding measure of one sixth of a circle, thus: If 90 gives 6, what will 60 give? Answer, 4; and 4 added to 108 is equal to 112, the area; or, as the circumference is 38 inches, and the hexagon, or 6 times the radius, measures but 36, you will see the circle gains 2 inches on a line; but as the line has no area, I square it, thus: $2 \times 2 = 4$; and this added to 108 makes 112, which proves that all give the same area if the diameter be 12 inches; and it also proves the circumference right, for as it gains just 4 inches, there is no other measure that would perfect the measure of the area; so the one proves the other.

Now, to find the triangle equal to the area, as the six sides of the hexagon measure 36 inches, it wants $\frac{1}{18}$, equal to 2, which added to 36, make 38, the measure of the circle. But the circle being a curved line, it loses $\frac{1}{3}$ of $\frac{1}{19}$, equal to $\frac{1}{57}$, of a straight line. So half the circumference being 114 sixths of an inch, subtract $\frac{1}{57}$, equal 2, from 114, which leaves 112, equal $18\frac{4}{6}$ for the base of the triangle, and the erect side is the radius, equal 6 inches, and the side wanting is the hypotenuse; but instead of seeking it, the best way is to take the base line, 112, with the erect line, 36, equal to 6 inches, or radius, and there being two triangles, reverse them, and they will make an oblong square, $18\frac{4}{6}$ inches by 6; multiply it together, and it will be the area. Thus:—

$$\begin{array}{r} 18\frac{4}{6} \\ 6 \\ \hline 112 \end{array}$$

112, perfect area.

Or you may take one half the circumference, equal 114, and subtract $\frac{1}{57}$, which leaves 112 sixths of an inch; it makes an oblong square, $18\frac{4}{6}$ by 6, the radius; multiply it, which makes 112, area. If you do not subtract $\frac{1}{57}$, you will find when you come to straighten the circle, to make the square, it would expand, and make the area too much.

The question is asked, Can you square the circle? or, what is the square of the circle? Answer: I take the hexagon, 12 inches in diameter, and divide every angle into two equal parts, making it 12 sides, or a dodecagon, which will measure just 108 inches, as may be proved by the oblong square, as shown above.

So the dodecagon is no part of a circle. Now, the six sides of the hexagon measure just 36 inches, and it gains just 2 inches in making it into a circle; therefore the circumference is 38 inches, and the two that are gained make up the measure of the circle. So you see the square of 2 is the square of the circle, thus: $2 \times 2 = 4$.

So 4 is the square, or 4 is the area of the circle; and adding the area of the dodecagon, equal 108, to 4, makes 112, the area of the round table; but to measure a circle, as people generally do, the same as square measure, from an angle of 90 degrees, it would make 6 instead of 4 for the square of the circle.

But if people understood how to work trigonometry on a circle, it would prove but 4, or by the rule of three; if 90 give 6, what will 60 give? Answer, 4.

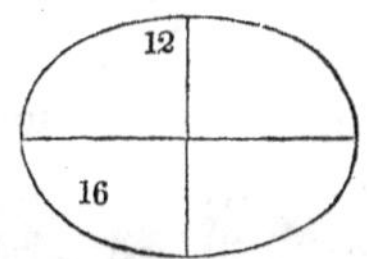

To find the measure of a true oval: Suppose the conjugate be 12 inches, and the transverse 16; multiply 16 by 12; then divide the product by 9, and $\frac{7}{9}$ will be the area.

To find the length to go round the oval: Take the measure of the conjugate, equal to 12, and multiply it by itself, thus: $12 \times 12 = 144$. Then multiply the transverse, equal to 16, by itself, and the product is 256; subtract the product of the conjugate, equal 144, from it, which leaves 112; extract the square root of 112, equal 10.583; divide the root by 2, which gives 5.292. This is the erect leg. Then get the circumference of a circle 12 inches in diameter, which is 38 inches, and divide it by 4, which gives 9.5 for the base; now square the two legs, and the square root of the sum will be one fourth of the circumference, which if multiplied by 4 gives the whole circumference.

To show that the measure to go round a circle is just $\frac{1}{57}$ part longer than a straight rod which measures the same area on the oblong square: Take 12 for a diameter; 3 times 12 are 36, for the six sides of the hexagon, commonly called the circumference. The radius is 6; so I multiply 36 by one half the radius, equal to 3, or one half of 36, equal to 18, by 6; either way gives the area of the oblong square, 18 by 6, equal to 108, which is just the measure of the dodecagon, or six oblong squares, 6 by 3, as shown in the plate.

Now for the measure of the circle. Find the circumference by the ratio, 9.5, which is just 38, gaining $\frac{1}{18}$, equal to square measure of 90 degrees. For example: Take a straight strip 38 inches long, for circumference of the longest ring, gaining 2 inches, which is $\frac{1}{18}$. Then set a bevel of 90 degrees; divide the 2 gained into 12 equal parts, making 2 inches, square measure, from 90 degrees. Then alter the bevel to 60 degrees, and mark it off just at the same distance as you did at 90 degrees, and you will find you lose just one third of your measure. Now, 60 degrees make the only equal triangle that can be made, and is the right angle to measure a circle, as one of 90 degrees is right for a square. Therefore, if 90 gives $\frac{1}{18}$, what will 60 give? Answer, $\frac{1}{27}$.

So 108 gains $\frac{1}{18}$ on the square of 90 degrees, equal to 6 inches, which gives 114, area; but at 60 degrees it gains only $\frac{1}{27}$, equal to 4 inches; so if 90 gives 6, 60 gives 4; and 4 added to 108 makes 112, the true area. Now, 112 makes an oblong square just $18\frac{4}{6}$ by 6; so you see it takes $\frac{1}{57}$ more to measure a circle than it does to measure the two sides of the oblong square, for the oblong square is straight measure.

Proof: As the circle gains 2, and the square of 2 is 4, which is the square of the circle, so the 108 gains $\frac{1}{27}$, equal to 4, making 112, area. So you see the area proves the circle, and the circle proves the area to be perfect measure.

On a former measurement of the circle that has come under my observation, the circle being 12 inches in diameter, they called the circumference $37\frac{7}{10}$. Now, suppose this to be the right measure; then three times the diameter, 36, would make an oblong square 18 by 6, equal 108 inches. Take 3 times 12, equal to 36, from $37\frac{7}{10}$, which leaves 1 inch and $\frac{7}{10}$, which is what it gains by being brought to a circle. So I square it, thus:—

17
17
———
289

So you may see that 2 inches and 89 hundredths is the square of the circle; I then add the square of the circle to the oblong square, thus:—

108
2.89
———
110.89, area.

But if you add the angle that it gains, and work it as 90 degrees, equal square measure, it gains $5\frac{1}{10}$, which added to 108 makes $113\frac{1}{10}$, area. By this you add the triangle of square measure, instead of the square of the circle; but if they had brought 38 circumference, its area would be 114, by their way of working it.

The great circle of this earth is what measure is derived from. Long measure, square measure, cubic measure, dry measure, and all measure, is to be derived from an equal proportion of this great circle, of and in which we exist; and unless that equal proportion is found, no measure is or can be correct; and that equal proportion must be found by a perfect measure of the great circle. Now, as an inch was calculated to be derived from an equal proportion of the great circle, and this derivation was from the proportion of as 7 to 22, the inch is imperfect, unless the proportion as 7 to 22 is right; and if the inch is imperfect, the foot, the yard, the rod, the mile, the league, and all other space and distance that is calculated to be derived from proportion of equality, must be also imperfect.

The proportion of equality that is derived from as 7 to 22 is imperfect, for the reason that the circle will not prove the area, nor the area prove the circle; neither will the circle prove the square, nor the square prove the circle; nor will the square prove the area. Now, 1728 cubic inches of water are calculated to be one foot; and according to specific gravity, the foot of water weighs $62\frac{1}{2}$ pounds. From this pounds and ounces are calculated; an ounce is a certain number of inches of this water; and the correctness of pounds and ounces must be derived from an equal proportion of the circle, the same as measurement.

This proportion, as 7 to 22, is so imperfect in the operation of mathematics, that in astronomical calculations it makes mathematics useless. Thus it has been considered, since the present operation of mathematics has been known; and instead of using mathematics in astronomical calculations, observatories have been erected, to observe distance, and space, and equality, for purposes to benefit mankind, which mathematics would not perform with correctness, in its present state of imperfection, which imperfection has been for the want of a perfect quadrature of the circle. These observations are attended with great expense, when compared with mathematics, with a perfect measure of the circle.

The measure of the circle shows that if you travel on a radius of one hundred miles you are $5\frac{1}{2}$ miles astray, for the want of perfect measure. Also, in the traverse of the earth round the sun, the calculations have been 1,500,000 miles in 24 hours; whereas correct measure makes it 1,700,000. Also, it shows that no level can be found on the earth, for the want of a perfect measure. It shows that the survey between the two oceans, at the Isthmus of Darien, was very imperfect, for the want of the perfect measure of the circle.

I can see no necessity for altering the names applied in long measure, square measure, dry measure, or in any measure, except the inch. A mathematical inch is the thirty-eighth part of a circle 12 inches in diameter. When the inch is made perfect, the foot, the yard, the rod, the degree, the league, &c., &c., are all correct.

The fundamental principles and bases upon which mathematics is built have slept in oblivion since the world began. Space and distance, with equality with the great circle of the earth, have never been known by any mortal man. That equality of the great circle of this earth, that rectifies all weights and measures, which is a great cause of quietude, peace, and happiness with mankind, has slept in oblivion, incomprehensible to man. It has been the strife and anxiety of all the most learned and the lovers of science, since the commencement of the world; large sums of money, and much time and attention, have been made use of by all nations to effect a perfect measure of the circle.

John Q. Adams's report of 1821, on coins, weights, and measures, will give a perfect history of the great exertions, strife, and anxiety with all nations for this measure. Mr. Adams says it has been the object of all statute making to guard against fraud and oppression, in consequence of the imperfection of weights and measures.

Space and distance can now be equalized by some measure of invariable length as a standard, which has been hitherto sought. Every man, woman, and child can now be equal in weight and measure, which is joy to their souls and peace to their minds.

The prevailing opinion in the United States is, that we have a standard of weights and measures; that no unrighteousness is done in mete or yard; a just balance, a just weight, a just ephah, a just hin, is measured and weighed to all.

We have in the United States some 48 custom houses. One house uses for a bushel 1925 inches, another, 2063, another, 2216, another, 2203, another, 2158, another, 2358; and in this way does it vary, making the average variation in all 75 inches in a bushel; and so it is all over the world.

This measure is now comprehended; equality, distance, and space can now be measured to perfection; all weights and measures, gold and silver coin, can now be determined, to the satisfaction of every man, woman, and child. That law, not by human legislation, but by God, on Mount Sinai, can now be fulfilled; a just balance, a just weight, a just ephah, a just hin, can now be made to all the world.

Having the circumference, to find the diameter: Divide the circumference by 19, and multiply the product by 6, which gives the diameter.

The hexagon, with the number 6, is what measures the circle.

It is the fourth power, or biquadrate, which makes up the measure of the circle.

It is what is contained beyond the dodecagon; it is $\frac{1}{18}$ in the circle, and $\frac{1}{27}$ in its area.

The circle is perfectly synonymous with the square.

The square of the circle may be considered by some to be the square that 12 diameter in a circle would make, which would be $9\frac{1}{2}$ inches.

I consider the square of the circle to be the square of the square, which is the square of 2, thus: $2 \times 2 = 4$.

Square measure is from an angle of 90 degrees; circular measure is from an angle of 60 degrees.

Circular measure gains over all other measure; once in diameter is four times in area.

To measure a globe or sphere: Suppose 12 is the diameter; first find the circumference, which is 38 inches; then multiply it by the diameter, 12; $12 \times 38 = 456$; divide 456 by 19, which gives 24; add the 24 to 456, equal 480 inches, the surface of a globe 12 inches in diameter. The surface of any globe may be found in the same way.

To find the solid contents of the same: First find the square of the diameter, thus: $12 \times 12 = 144$; multiply 144 by 12, which gives 1748, cube; divide the cube by 12, which gives 144; multiply 144 by 7, which gives $\frac{7}{12}$, or 1008, the solid contents of a globe 12 inches in diameter. Any globe may be measured in this way.

For example: Suppose a globe 6 inches in diameter:—

```
           19, circumference.
            6
          ---
     19 ) 114 ( 6
          114
          ---
            0

Add the 6 to 114
               6
             ---
             120, surface.
```

Surface twice in diameter gives 4 times surface. Globe twice in diameter gives 8 times as much solid contents.

```
           6, solid contents.
           6
          --
          36
           6
         ---
   12 ) 216, cube.
         ---
          18
           7
         ---
         126, solid contents of a globe 6
                  inches in diameter.
```

To find any one side of a right angled triangle, having the other two sides: In every right angled triangle the square of the hypotenuse is equal to both the squares of the two legs. Therefore, to find the hypotenuse add the square of the two legs together, and extract the square root of the sum. And to find the leg, subtract the square of the other leg from the square of the hypotenuse, and the square root of the difference is the leg required.

Example 1. What is the hypotenuse of a right angled triangle whose base is 56, and perpendicular 33?

```
   56          33
   56          33
  ----        ---
  336          99
 280          99
 ----        ----
 3136        1089
 1089
 ----
 4225 ( 65 feet, answer.
```

Example 2. What is the perpendicular of a right angled triangle whose base is 40 yards, and hypotenuse 50 yards?

```
   40          50
   40          50
 ----        ----
 1600        2500
             1600
             ----
              900 yards, answer.
```

Example 3. What is the height of a scaling ladder to reach the top of a wall 28 feet in height, and across a ditch 45 feet in breadth? Answer, 50 feet.

Questions of this nature are resolved by the foregoing problem; the height of the ladder being considered as the hypotenuse of a right angled triangle, and the height of the wall and breadth of the ditch as the two other legs of the triangle.

CIRCULAR ROOT.

To find the hypotenuse on a circle, having the diameter of any circle given: First find the circumference, and take any part of the circle, or the whole of the circumference, as needs may require, and call it the base, and square the same; then take the erect leg, and square it also, and add the two sums together, as in extracting the square root; and the square root of the sum will be the circular hypotenuse.

By this rule you may find the length of a circular stair rail, or any thing that may be wanting of that circular form, just the same as the square root.

Example: Suppose a circular flight of stairs 12 feet high and 6 feet in diameter, equal 19 feet base; what is the length of the hand rail?

```
            19                 12
            19                 12
           ----               ----
           171                144, square of leg.
           19
           ----
           361, square of base.
           144
           ----
        √  505 ( 22.472
           4
           ----
      42 ) 105
            84
           -----
     444 ) 2100
           1776
           ------
    4487 ) 32400
           31409
           -------
   44942 ) 99100
           89884
           -------
               0
```

Answer: the stair rail is 22 feet and 472 thousandths of a foot long.

Space and distance, with equality of the great circle of this earth, have never been found; consequently, no man knows what an inch is, a foot, a yard, a rod, a league, a mile, a degree, a pound, an ounce, a glass, a gill, a half pint, a pint, a quart, a gallon, a peck, a half bushel, a bushel.

These measures and weights enter the concerns, privately, individually, domestically, of each and every individual upon the earth, young and old, rich and poor, great and small, from the king to the beggar. Equality and space are the fundamental principles and bases of moral justification of right and wrong; without these no man knows whether he is rightly used or wrongfully abused. They are the mediators of contentment and justice, and teach mankind to do right. They enable men to have confidence and quietude in all their transactions with each other. They are the mediators between the honest man and the knave, and steps to salvation that make glad the heart of man.

It is the supposition and belief of the majority of mankind that we have a standard of weights and measures throughout the world; but this is erroneous; there is no standard of weights and measures, because such a standard would be uniform, and such a thing as uniformity of weights and measures is not known in this world. For example, England may have what she calls a standard, but this standard will vary in every state and kingdom, town and county; no two are alike, and there is no one that is right. How can it be possible to make a correct standard of weights and measures, when that correctness proceeds from a perfect equality of the circle of the earth, and that equality has never been known by any living man or mortal since the world began? That equality has been incomprehensible to man; therefore justice and honor to all men, as regards weights and measures, has been wavering to the four cardinal points. If a man gets a pint, quart, or gallon, it is well, and if not, all the same; for who is to say whether it is mete or measure? As figures are inexpressible to man, he cannot determine its equality. To obtain this measure has been the continual exertion of England, France, and Spain, for hundreds and thousands of years. At one period France kept twenty or thirty of her most scientific and learned men for seven years, endeavoring to arrive at some determined equality of measure. The ambition of these men was so much aroused that they sailed round the world.

Thomas Jefferson says, the first object that presented itself was the discovery of some measure of invariable length, as a standard. There exists not in nature, he says, as far as has been hitherto observed, a single subject, or species of subject, accessible to man, that presents a uniform dimension.

OBSERVATIONS.

I have carefully collected these ideas, thinking they might be of some use to my fellow-men; and if I have accomplished my desire to the satisfaction of my readers, my prayers are answered; and if not, it cannot be imputed to the want of an honest intention.

The brute creation may perhaps enjoy the faculty of beholding visible objects with a more penetrating eye than ourselves; but spiritual objects are as far out of their sight as though they had no being. Nearest to the brute creation are those men who suffer themselves to be so far governed by external objects as to believe nothing but what they can comprehend with their shallow and imaginary understandings, and who let such expressions as these fall from their lips, in floods of abundance: — "O, it is impossible — it cannot be — no reasonable man can think so!" as was said to me, when endeavoring to convince them of my discovery. I was told that no rational mathematician would ever make the attempt. Proud, vain, and foolish man! how unwise art thou, for the want of that good reasoning which Jesus Christ dictated to the world! Let us all gain wisdom by moderate, cool, thoughtful closet prayer; consult in private with your heavenly Father, for he is your God, and not man. Let all your meditations be with him, and not with man. Your conscience is the mediator between yourself and God. This putting trust in and worshipping men and idols may be imputed to the representation made use of regarding beasts, as confined to earthly objects, and not to our heavenly Father, who seeth all things on earth as well as in heaven.

If I have done myself a credit, I hope it may be imputed as an honor to man; I wish not to boast, as it is not of me. Friendship, harmony, love, universal kindness to all men, and a fervent desire for peace and contentment after this life, are the heart throbbings and conscientious feelings of your humble servant.

TESTIMONIALS.

I have travelled with Mr. John Davis, in England and the United States, for six years; we have visited all the most learned mathematicians that we could hear of, but have never found one that attempted a disapproval of his work. I have examined, to the best of my ability, the works of writers on mathematics since the time of Euclid; and without a doubt, according to mathematics, he has solved the wonderful problem, surprising as it may appear to many.

Weights and measures can now be made perfect, and to the satisfaction of every man, woman, and child. Space and distance, which have never been known, are now found by John Davis.

SABIN SMITH,
Attorney and Agent for John Davis.

Nothing is more recreating or interesting to a mathematical mind than John Davis's measure of the circle. It has been the study and strife of our learned and scientific men since the time of Archimedes; and it is a manifestation of supremacy that this measure is perfected. Its value pen or tongue cannot describe. I anticipate great improvements from this measure in the next fifty years, to the pride, happiness, and benefit of mankind. No doubts can rest with me as to the perfection of this measure.

THOMAS ROSSITER.

I have had an opportunity for two years to be with Mr. Davis, and during this time I have given great attention to his work — the measure of the circle. It has been my business, since my manhood, and always was and still is my pride, to practise in mathematical operations, such as surveying, &c. I have had transactions with many able mathematicians, but have never met with such a man as John Davis. He has a mind which exceeds, in the comprehension of mathematics, all others with which I am acquainted. His measure of the circle is a wonderful discovery, and no doubts can be entertained of its perfection, when mathematically demonstrated.

Asahel Buck.

I have wrought as a mechanic, at the tin ware business, for fifteen years. I became acquainted with Mr. Davis in New York city, while working at my business. Mr. Davis and Mr. Smith taught me the measure of the circle, so that I could find the circumference of a circle, and its area. I have made use of his measure in my work, and find it perfect.

Charles Wilkinson.

New York, July 17, 1850.

I have taken into consideration the measurement of the circle, and have examined it to the best of my ability, giving it much time and attention; and do say that, to the best of my belief, a complete measurement has been effected by Mr. John Davis.

M. F. Greenlaw.

I have examined the measure of the circle by Mr. John Davis, and find it, in my opinion, a most complete, scientific, mathematical measure of the circle.

Thomas Guile,
Professor of Mathematics.

I have wrought as a mechanic for twenty years, and in some of my mechanical operations I have found it very difficult to match my work from the proportion of as 7 to 22, and by experimental operations I came to the measure of three times the diameter, and one sixth, and from this I have found no difficulty in matching my work; and when Mr. Davis told me that three and one sixth times the diameter was his proportion, I was satisfied that his measure was correct.

A Mechanic of Paterson, N. J.

I have examined the measure of the circle by John Davis, and beyond all doubt it is perfect measure.

Henry R. Savory,
Civil and Military Engineer.

"THE CIRCLE OF THE FAIR."

BY THOMAS ROSSITER.

We all have had a tangle
  In the "*circle of the fair*,"
And stricken by a charm
  In our bosom — funny, queer: —
But who can straightly tell us,
  Independent of a doubt,
Once in the blessed circle,
  How the deuce they did get out?

Remembering the spider's
  Habitation for a fly,
We're fearful to encounter,
  Yet seldom fail to try;
For love is more enticing
  Than the pearls of a sham,
And seldom fails to better
  Or subdue the inward man.

Once, in the Eagle valley,
  By the Hudson's gentle tide,
I saw a faithful shepherd
  Press a maiden to his side;
She was telling how she loved him,
  And how heightened was her bliss;
Though the flock strayed on the wolf-path,
  He was kissing her for this.

Now, if we are unlucky
  In the "*circle of the fair*,"
We need but welcome Davis,
  He can for us make it square;
So drink a health to Davis,
  While the goddess of our theme
Shall raise the wreath of Flora,
  As an emblem of his fame.

www.ingramcontent.com/pod-product-compliance
Lightning Source LLC
LaVergne TN
LVHW021407110826
845150LV00007B/1814

9781425512460